Principles of Phonometrics

ALABAMA LINGUISTICS & PHILOLOGICAL SERIES No. 18

Eberhard Zwirner and Kurt Zwirner

PRINCIPLES OF PHONOMETRICS

translated
by
H. Bluhme

The University of Alabama Press
University, Alabama

Translated into English from
Grundfragen der Phonometrie Copyright © 1966
by S. Karger AG Basel
English translation and addenda Copyright © 1970
by The University of Alabama Press

Standard Book Number: 8173–0153–4
Library of Congress Catalog Card Number: 70104932
Manufactured in the United States of America

Translator's Preface

Considerable time has elapsed since the first edition of *Principles of Phonometrics* was published in German. It may, therefore, be seasonable to point out the inherent relevance of this treatise to some philosophical aspects of linguistics. The joint authors, Eberhard Zwirner and Kurt Zwirner, document the development of their linguistic thought with a wealth of historical material and facts; it should, however, be noted that the work under consideration is in no way intended as a manual of the history of linguistics. It bypasses the bulk of Greek and Indian contributions to the advancement of linguistic methods. The omission, particularly of Panini, may perhaps be regretted, but a sufficient number of handbooks and journal articles are available to fill this gap. Also, Hebrew-Arabic scholarship and the Chinese phonetic treatment of versification are not represented in this volume—the reason being that the historical significance of these ideas for generating the thoughts put forth here is ambiguous.

The authors first sketch a brief outline of phonometrics. This prelude is followed by some notes on Greek contributions to linguistic thinking. The subsequent treatment of the renaissance emphasizes Galileo's contribution to the philosophical concept of the scientific method. On the surface, this part may seem to be irrelevant to phonetics, but the authors clearly demonstrate how these ideas furnish the basis of their view of the function of phonometrics; namely, as linking the concept of the linguistic unit to its physical, measurable aspect.

In particular, effort is devoted to demonstrating the influence of 19th century achievements in the natural sciences on the early school of comparative linguistics. Historical and biographical details are mentioned that throw new light on linguistics from Bopp to Trubetzkoy, stressing the close relationship between the different disciplines as well as the point where they methodologically diverge.

The amount of historical information in what is basically a book on the philosophical foundation of phonetic research may be surprising, but this knowledge very well serves the purpose of giving new perspective to hitherto little elucidated developments in European linguistic thinking of the last two hundred years. The dominant role played by 19th century German linguistics explains the authors' heavy reliance on sources which seem to be largely unknown outside the European continent, as could be said of the philosophical trends which become apparent in Zwirner's work. It should be noted in this context that E. Zwirner is a leading expert on Neokantian philosophy as expressed by the Marburg school of thought. His contribution to a new scientific concept of the methods and problems of phonetic research, as a branch of the system formed by all sciences, is equal to that which Richard Hönigswald and E. Cassirer have achieved in a broader linguistic framework. This Kantian theory of cognition forms the cornerstone of the phonometric method.

The clarification of this method—the epistemological problem of sound segmentation and identification, the question of the formation of sound classes (roughly equivalent to the phoneme), and the relationship between the phoneme (sound class) and its physical realisation in the phone—forms the essence of the second half of this volume and is dealt with both on the physiological and acoustic levels. The difficulty of the relationship between phoneme and phone is resolved by the principle of coordination, in which the variations of the phones are associated with the fundamental invariable of the phoneme. This reference of the phone to the phoneme is determined by the phoneme's zone of statistical phonetic variation.

The detailed presentation of the execution of phonometric procedures is supported by references to general statistical examples. The phonometric terminology may be somewhat isolated from current trends, but the translator has been primarily concerned with delineating, precisely the intricate philosophical and stylistic paths of the authors.

Herman Bluhme

Author's Preface
to the English Translation

Modern linguistics differs from 19th century linguistic research by its clear distinction between synchronic and diachronic aspects of language and its emphasis on the synchronic patterns prevailing at any historical stage of a language. It also maintains a distinction between Saussure's "linguistics of speaking" and his "linguistics of language."

The principal task of modern linguistics, as applied to individual living languages, is to detect and verify the realization of these synchronic patterns in utterances.

Consequently, modern phonetics has to be "linguistic phonetics: experimental phonology or phonometry," as Hjelmslev chose to define it some thirty years ago. Nowadays, phonetics can no longer be regarded as the study of speech sounds independent of linguistics; it has become the study of the *spoken* language and forms an integral part of linguistic research. Phonetics is the scientific investigation of the realizations of the synchronic structural patterns used in the speech community concerned. In this sense, it is termed functional phonetics.

Long before the technical means for acoustic and physiologic research were available, a number of distinguished linguists—Rudolf von Raumer, Wilhelm Scherer, Georg von der Gabelentz, and Ferdinand de Saussure, among others—had pointed out that the proper object of study in scientific linguistics is not the written text but the spoken language itself. As spoken, a language exhibits its own inherent patterns of semantic, lexicologic, syntactic, morphologic, and phonemic structure, all of which are subject to variations in their actual realization. Hence, the description and analysis of spoken language must be considered an integral part of linguistics.

Today, the linguistic approach that distinguishes between the structure of language and its realization in utterances has made the investigation of spoken languages feasible. The invention of various recording devices—record players, tape-recorders, high-speed photography and cinematography—and the discovery of procedures for the linguistic evaluation of the language material have prepared the ground for this development.

This book represents an introduction to the history of these investigations, to the principles underlying functional phonetics, and to the techniques of "experimental phonology."

The first edition was published by Metten & Co., Berlin, in 1936; a second enlarged edition, by S. Karger Publishers, Basel/New York, in 1966—both under the title *Grundfragen der Phonometrie*.

Recording techniques and methods of the scientific description and analysis of tape-recordings, high-speed motion pictures, and cinematographic data have been greatly perfected in the past thirty years, but the basic problems involved in the analysis of the acoustic signal and the registration of the physiologic processes have remained unchanged. These problems will doubtless continue to exist in any quantitative analysis of tape-recordings. In view of this, it seems to me to be permissible to publish a translation of a book thirty-four years after it first appeared in 1936.

I wish to thank Dr. Bluhme for his preparation of the English version and the Translation Unit of the Australian National University for its assistance. I also wish to thank Doctors M. Clyne and K. Gardiner for proofreading. I am grateful to the University of Alabama Press for publishing the book.

Münster/Westphalia   Eberhard Zwirner

Table of Contents

Translator's Preface   v
Author's Preface to the English Translation   vii
  I.   The Task of Phonometrics   3
 II.   Observations on the History of Phonetics   8
     1. History of the Term 'Phonetics'   8
     2. Ancient Phonetics   15
     3. Renaissance Phonetics   18
     4. Phonetics of the Seventeenth Century   26
     5. Phonetics of the Eighteenth Century   30
     6. Phonetics of the Nineteenth Century   53
III.   Methodological Foundations of Phonometrics   82
     1. Two Objections to the Distinction and Distribution
        of Linguistic Segments   82
        Syntagmatic Objection   83
        Paradigmatic Objection   87
     2. Sound Class and Sound Realization   89
     3. The Sound Unit (Sound Segment)   94
        Linguistic Unit of Sound   94
        Auditory Sound-Unit   99
        Physiological and Acoustic Sound-Unit   105
 IV.   Phonometric View of the Sound System   122
     1. System of Distinctions and Oppositions   122
     2. System of Segments and of Variants   127
  V.   Execution of Phonometric Procedures   131
     1. Compilation of Transcribed Texts and Phonometric
        Text Lists   131
     2. Variation of Phonometric Features   150
Notes   157
Bibliography   175
Index of Persons   190

Principles of Phonometrics

I. The Task of Phonometrics

We shall argue that the various approaches covered by the term phonetics postulate, so far as they follow scientific aims, a system of languages and dialects and that their aim is to clarify this system. Further,
we shall show that, so far as phonetics has contributed to the investigation of languages and dialects, its measurements and quantitative
procedures—here summed up by the term phonometrics—also have linguistic aims, that these are not the same aims as those of the natural
sciences, and that phonometric procedures are therefore part of linguistics. This is what we intend to show in this book and in further
articles through the discussion of the problems and objectives of phonometrics and by presenting the results of phonometric investigations.

When we investigate laughter, crying, or shrieking, we deal with
autonomous physiological problems because they are hereditary characteristics, whereas speech has to be learned on the basis of inherited
aptitudes. In singing, we are concerned primarily with problems relevant to musical history. Just as there is no such thing as "speaking in
general" that would be valid for all people distinguished only by racial
characteristics, "singing in general" does not exist. However, the way
in which singing is transmitted from generation to generation is essentially different from the intergenerational transmission of language,
although occasionally boundaries are vague. This becomes very evident when we realize that linguists never find it necessary to deal with
historical changes in the field of music—or more specifically of singing.
Even in the rare cases in which sounds that are sung serve to convey
meaning, and in which it may be impossible to decide without making
an artificial distinction whether we have to deal with speaking or singing, these sounds are only of interest (for example, to scholars of Afri-
can languages) because here they have taken over speech functions.
In relation to these, the various functions of singing become irrelevant.

The physiological and physical problems of singing are therefore relevant to the diachronically orientated approaches in phonometrics only so far as they further the investigation of languages.

Renouncing the allegedly new research methods of the "natural sciences" in quantitative phonetics, we most definitely want to return to the linguistic objectives of the last century. We also wish to follow the structural endeavors of the last decades in the belief that a comprehensive knowledge of the structures valid within the fields of geography, sociology, and history is an essential prerequisite for all comparative investigations in these fields. Such considerations rest, to be sure, on the assumption that there are large unexplored fields of language that may be familiar to the ear but that cannot be investigated without measurements and statistics.

It seems that the opinions of the Neogrammarians have been abandoned too hastily and without sufficient verification. It is likely, as they suggested, that fluctuations of phonemic-allophonic realizations, which may remain undetected by speaker and listener within a language community, actually play a significant role in sound changes. Certainly some changes occur that are not caused by such fluctuations; for example, when new words or whole phrases are borrowed or newly coined. But Neogrammarians never claimed that the *gradual* changes they have postulated were the only principles of linguistic change. They merely proposed that sudden linguistic changes would often be only simulated by sudden changes in the writing, while in fact the writing system did not show these alterations until long after they had actually been accepted. This problem has been discussed frequently in Germany, at least since Rudolf von Raumer.

Evidence of fluctuations in the realization of phonemes has been sought for more than a century. The finding of such evidence has become a task of quantitative phonetics, as has the attempt to discover a geographical or social displacement of these variations. Such variations would then have to be postulated in historical linguistics as well, even if the evidence for them is obtained only after considerably more time will have elapsed following the development of high-quality sound recording.

Phonometrics is based on the assumption that the description and the synchronic or diachronic comparison of languages can be extended beyond what would be possible on the basis of the mere mastery of languages by speaking, listening, or writing. The different speech sounds reach our ears by physical processes, and sophisticated measuring instruments are available to register and take possession of these linguistic fields. Hermann Usener said (in 1882) that the philologist should not evade any problem even if "it had to be solved with ruler or balance, by calculations, or with geometric constructions." This applies also to the linguist.

The following investigations will be limited to the study of those relations which can be assessed by measuring and counting and which are suitable for describing and distinguishing languages. The reason for such a limitation may become clear when we compare our task with the task of phonetics, which occupies itself not only with scientific problems but also with problems of pedagogics, phonetic transcription, orthography, shorthand, voice formation, and singing; while its scientific aims are divided between linguistic and physiological objectives as well as between psychological and physical ones. Phonetics likes to be considered an interdisciplinary study simply because it is—in the Kantian sense—not oriented towards a method; that is to say, towards a clearly defined scientific task. Instead, it is determined by an object that no other science defines. It considers that all endeavors related to this object lie within its own field.

However, at first glance, phonometrics also is not uniform—let alone a clearly defined field—containing, as it does, such aspects as preliminary studies of sound statistics and the analysis of the vibrations of spoken sounds. The analysis of sounds that are uttered in the context of discourse must be distinguished from the analysis of the vibrations of vowels that are sung—a science that has reached such a high degree of perfection through the investigations starting with Leonard Euler, Robert Willis, and Chladni; continued by Seebeck, Ohm, and Fourier; and then by Helmholtz and Hermann, Stumpf, Wagner, and F. Trendelenburg. In addition, phonometrics deals with the measurement of sound length, pitch, and stress, and the measurement of the positions and movements of the articulatory organs.

In all these cases, our methods are measurements and quantitative procedures. But we should not think that these procedures constitute language or communication; they are, however, in all instances *associated* with speech. At each stage, these quantitative procedures remain dependent on linguistic concepts and methods. No attempt is made to substitute measurements for linguistic investigations, because languages themselves cannot be measured. Instead, the measurements and calculations should serve to supplement the results and methods of linguistics.

These measurements and calculations, however, are not in themselves of the same nature. Even if they were all ultimately used in linguistic research, it would remain necessary to realize that the common aim is achieved in different ways. The perception of language, the counting of sounds, the measurements of sound vibrations, and the measurements of the movements of the organs of speech all relate to phenomena that are part of language; nevertheless, they differ among themselves about as much as psychology, mathematics, physics, and physiology differ.

In the perception of sounds, we have a situation in which a person, calling himself "I," hears something and understands it by referring it to a traditional system of sounds with which he is familiar; in short, we have a situation in which the postulates of psychology cannot be ignored. In the counting of sounds, we utilize statistical statements and calculations of frequency or of the *predominant usage* of linguistic phenomena. This cannot be accomplished accurately without the tools provided by mathematics. In the measurement of sound vibrations, we analyze oscillations in the density of air pertaining to the perception of speech sounds. These oscillations cannot be represented without the tools of physics. In describing the activities of the human speech organs, it is necessary to have recourse to the methods of anatomy and physiology. None of the speech organs is used exclusively for speech; their activities are only of interest so far as they serve in communication.

These diverse quantitative methods stand in the same relation to the common goal of linguistics as biometrics stands to the investigation and comparison of organisms. For this reason, these methods are summarized in the term *phonometrics*. To put it in another way, phonometrics relates to linguistics as speech relates to language. Phonometrics cannot be more, and does not profess to be more, than the totality of all the technical methods of the investigation of speech that serve linguistics: it provides for linguistic methods of measuring and counting suitable for the investigation of language and speech. Phonometrics is based on the assumption that language, as the sum of variable and varying structures, is the product of an historical process. Phonometrics also presupposes that the empirical and perceptible realization of actual speech or conversation is determined by linguistic as well as psychological, biological, physical, and mathematical factors. The task of phonometrics is to bring all these methods to bear on meaningful speech and to coordinate the results obtained by all these different approaches so that they may serve the purposes of linguistics. A scientific method that would rest on postulations peculiar to phonetics does not exist.

In the present work, we shall deal with the question of the *unit of speech sounds*; specifically, of the sounds that are uttered in the context of discourse and conversation, because it is in this direction that crucial problems lie hidden. We refer to the problems of counting and totalling speech sounds and of justifying phonetic transcriptions (that is, the utilization of a finite set of phonetic symbols), and finally to those problems relating to the interpretation of graphs that represent air molecules brought into oscillation by speaking, thus showing the movements of the speech organs in the act of speaking. The theoretical solution to all these tasks lies in grasping the relations between language and speech.

The verification of the concept of the unit of speech sounds would provide the foundation for a systematic science of language statistics, and indeed could serve as a basis for any kind of quantitative linguistics. In other words, it would provide the substructure for all the work which has been attemped by classical, experimental phonetics since Rousselot, so far as such investigations have followed genuinely scientific objectives.

The present work further aims at showing that any quantitative treatment of language and speech presupposes the fact that speech sounds can be counted. This applies also to those cases in which experimental phoneticians believed that, on the basis of their own enquiries, they had discarded the concept of speech sounds as units and hence refuted the justification of the symbolic representation of speech sounds by phonetic signs. The core of truth that lies in the arguments which the experimental phoneticians adduced against the divisibility of word perception and of their corresponding graphical representations must certainly be taken into consideration and its importance recognized. On the other hand, we must make it perfectly clear that two curves (for example, those of two plosives or two vowels) cannot be compared or related to each other until we have decided that they are the graphical representations of two sound *units*—whatever these may be. As soon as the concept of the linguistic units (which is simply another name for countability) is questioned—be it that of a sound, a syllable, a word, or a sentence—experimental phoneticians will become involved in insoluble conflict with linguists and will thereby see the basis of their position eroded. As it is possible to speak of language and communications only if we assume the unity of speech sounds, we must postulate that the quantitative treatment of speech rests on the same assumption.

II. Observations on the History of Phonetics

The difficulty of an historical exposition of phonometrics lies not only in the fact that many preliminary studies necessary to a synoptic treatment of the field are still lacking, but especially in the fact that the endeavors compressed into *one* task in phonometrics have developed almost independently of the rest, and in part exist side by side even today without closer academic contact having been established between them. In what follows, therefore, we are not attempting to present a synoptic exposition of all historical tendencies but merely to cite individual documents from the history of phonetics and the related sciences, quote them in part, and consider them critically, so far as they are significant for a current evaluation of the quantitative treatment of living languages.

1. HISTORY OF THE TERM 'PHONETICS'

Rousselot's[1] view, that phonetics owes its name to Baudry[2] and Bréal,[3] is based on an error. Apart from the Greek authors,[4] Georg Zoega was the first to use the word "phoneticus," in 1797 in an exposition of the hieroglyphic symbols:[5] "Sed satis est exemplorum classis aenigmaticae, super est quinta classis notarum phoneticarum, quam ad aenigmaticam referri posse jam monui, sed quoniam a plerisque neglectam invenio ad ea autem quae infra dicenda erunt de alphabeti origine maxima ejus est utilitas separate de ea agendum duxi." [6]

In 1822 J. F. Champollion adopted the concept of phonetic hieroglyphs and, with his deciphering of the hieroglyphs, introduced this idea into research once and for all: "Le monument de Rosette nous présente l'application de ce système auxiliaire d'écriture que nous avons appelé phonétique, c'est-à-dire exprimant les sons, dans les noms propres des rois Alexandre, Ptolémée, etc." [7]

In his discussion of Young and Salt,[8] who were placed in the right perspective by Russell in 1831,[9] the concepts of phonetics and of phonetic hieroglyphs are already used as established scientific terms.

In 1826 W. Kirby and W. Spence[10] attempted to introduce the word "phonetics" into entomology, but without success.

In 1833 Franz Bopp was already using the word "phonetisch" fairly frequently, and in a meaning for which he otherwise has "euphonisch": "I think I may regard the lengthening of the final vowel of the first member of such compounds as . . . not as purely phonetic, but as a consequence of the dual inflection";[11] "In the sibilant of this form I recognize no connection with either the character of the future, or the desiderative, but a purely phonetic addition."[12] Compare the latter quotation with the following: ". . . the σ in this compound is not a euphonic addition, but belongs to the stem . . . ."[13] It seems that in Bopp's writings the term "phonetic" does not appear alone, but always as "purely phonetic" or "merely phonetic."

Three years later, Wilhelm von Humboldt used the word "phonetic": "One may call the abstract of all the means which a language utilizes to attain its ends, its technique, and may divide this technique into phonetic and intellectual";[14] "I reserve the right to return subsequently to the superiority of technique generally. I only wished to refer here to what the phonetic may claim over the intellectual";[15] "when this goal is reached, the inner language development will not follow one-sided paths, on which it is abandoned by the phonetic production of forms, nor will the sound luxuriate in rich profusion over the fine requirements of thought."[16] An examination of Humboldt's works reveals clearly the origin of his use of the term "phonetic." In the Table of Contents,[17] the last chapter is entitled "The relation of writing to speech," and further: "—item. The phonetic hieroglyphs of Mr. Champollion the Younger." In 1856 A. F. Pott writes of an "addition . . . in a purely phonetic interest."[18]

In 1836 Friedrich Diez distinguished notations of sound that were based on a phonetic principle from those based on an etymological principle.[19] However, only the word was new in this connection, not the distinction itself. It already played a decisive role in all endeavors to promote the development and improvement of a unified German orthography and of a common German literary language. The connections of these efforts with the history of language and ideas has been outlined by Konrad Burdach:[20] Klopstock in the 1770's and 1780's had contrasted etymology and pronunciation. As one of the principles of his orthography, he recommended that no sound should have more than one phonetic symbol and no phonetic symbol should have more than one speech-sound.[21]

Diez modified the old distinction between the principle of etymology and that of pronunciation by his concept of the phonetic principle only so far as he shifted the emphasis, so to speak, from the speaker to the hearer. Thus the idea of community pertaining to language received full expression. Yet he restricted the new word, after the fashion of Zoega and Champollion, to the symbols which are to represent sounds. Rudolf von Raumer used it in the following year for the investigation of pronunciation itself. ". . . Since the conversion of words is not based on the written symbols," he writes in his famous dissertation of 1837,[22] "and on the resemblances of these, but on the spoken sounds, phonetic investigations must really go hand in hand with clear etymology." He designates phonetics as that which really matters to him.

It is quite possible that Raumer had an influence on the terminological changes in the 1840 edition of Grimm's *Grammar*. In Raumer's writing there is continual mention of phonetic nature, phonetic spelling, phonetic rules, a phonetic dictionary, phonetic validity, phonetic elements, and phonetic correspondence, so that one could say that he naturalized the concept of phonetic investigation in Germany—and indeed in the sense of a physiological definition of speech sounds besides the purely etymological one.

In 1838 Franz Delitzsch, by distinguishing a phonetic principle from a dynamic one, used the concept of phonetics in a new meaning again. The decisive passage is as follows:

> "Quin potius haec dispertitio literarum tantummodo historica via inveniri potest, h.e. collatione linguarum, observatione mutationis literarum et abstractione earum, quae cum certis ex legibus vicissim ponantur idque *phonetice*, non dynamice, naturam eandem participant eundemque ad ordinem sunt referendi." [23]

It appears that Delitzsch means here by "phonetic" a historical and etymological correspondence of the sounds, by "dynamic" their articulation.

K. M. Rapp's[24] work of 1836 is important for its attempt to present comparative grammar as a natural science: "The possibility of the history of language as a physiology of language is based on the identity of the human vocal organs throughout the whole race, from its creation onwards and in all zones. Its possibility from the logical point of view is based on the unity and community of the realm of thought. Language has therefore a natural history, a physiology, a physics, and a logical or historical development of concepts in it." [25] This one sentence suffices to show how, twenty years after Bopp's conjugational system and seventeen years after the first volume of Grimm's *Grammar*, Rapp had not yet grasped the meaning and the task of historical and comparative linguistics within the framework of his own physiological endeavor. On the contrary, he actually regarded it as the task of the century to transform linguistic history into natural history; that is, to describe language exhaustively by the means and conditions of natural science. Bindseil too, who moreover leans heavily on Kempelen and appeals to his authority, speaks in 1838 of the relationship of written to spoken language, of sounds, (phonetic) symbols and speech sounds, but does not employ the concept of phonetics for their physiology,[26] and similarly Johannes Müller in 1840 appears not to know or at least not to use it.[27]

It was not until 1841 that the German Wocher suggested the term "Phonetismus" for "organic sound formation," and "Phonetik" or "Phonologie" for its systematic investigation; here also the—at least theoretical—orientation towards the vocal organs is again embellished in a rather Humboldtian fashion: "The spirit [=*Geist*], which with feeling and imagination formed language, followed unconsciously and spontaneously the natural law which makes the spirit dependent on the simple arrangement of the human vocal organs." [28] Further on, he speaks of the secret weaving and forming of language, chiefly linked with the arrangement of speech organs.

In the same year Robert Gordon Latham[29] wrote an essay in which he proposed distinctions which remained, so far as we know, completely disregarded:[30]

How far the articulate sounds are systematically related to each other; how far and in what cases they run into each other; how far the chain of relationship is lineal, and how far it is circular; how far vowels and consonants, mutes and liquids differ in kind from each other; these questions, and questions similar to these, constitute the province of phonetics. [For Latham phonetics is the part of acoustics that is concerned with the relation and analogy of speech sounds. His term "acoustics" is to be understood as the auditory perception of acoustic phenomena—an attitude rather similar to Raumer's.]

These investigations must be distinguished, on the one hand, from those of physiology, and on the other hand, from those of etymology.— The science of phonetics determines how two given articulations are related; the science of physiology inquires how they are produced.— Phonetics tell us that between two articulations an immutable and essential relation exists; whilst etymology observes that under certain circumstances one articulation is changed into another. Very often etymology does more; it assumes the alliance from the change.—Now it does not follow, in etymology, that because in a given language one sound changes to another, those two sounds are, therefore, naturally allied to each other; although such is *often* (far, however, from always) the case.—Nor yet has it been proved in physiology that the sounds which to the ear sound alike, are produced by a like disposition of the parts of the mouth or larynx; although that such a correspondence exists is highly probable. The truth is that the special study of phonetics, instead of being promoted by the grammarian and anatomist, has been retarded by them. Etymological and physiological tests, etymological and physiological classifications, have been applied where acoustic principles alone ought to have been recognized. A necessary correspondence, moreover, between the three kinds of sciences, which in grammar may be proved nonexistent, and in physiology has not been proved at all, has for the most part been gratuitously assumed; and more than this, in those cases where the three tests have not coincided, the disposition has been to sacrifice the phonetic test to the other two.—The phonetic test stands thus—Two sounds are allied because, to the ear, they sound alike.—The physiological test thus—two sounds are allied because in the mouth and larynx, they are formed alike.—The etymological test thus—two sounds are allied because under certain circumstances, one is changed to the other. Expressed in Latin axioms, the etymological test is *propter fortunas articulatio est id quod est*, the physiological one is *propter formationem articulatio est id quod est*, and the phonetic one is *propter sonum articulatio est id quod est*. The coincidence of the three is—in the present state of our knowledge—to be considered as an accident.—Phonetic science considers sounds irrespective of the signs by which they are expressed, and irrespective of the names by which they are called. The sound of the *b* in *bat* is the same, whether it be expressed by the sign *b* or *β*; it is also the same whether it be called *bee* or *beta*. A sound is not double because it is spelt with two letters, nor yet single because it is expressed by one. The *th* in *thin* is a simple sound, irregularly expressed: the *x* in box is a double one compendiously spelt. In questions of the kind in point, the eye has so often misled the ear that the above given truisms are scarcely to be considered superfluous. In points of acoustics the blindest guides are the best.—To guard against the influence of names and signs it is convenient to express the relations of certain sounds arithmetically; and besides this—in the case of simple single sounds that have no simple single sign (or letter) to denote them—to coin or borrow appropriate signs as the occasion requires.

This was the clearest and most outspoken statement concerning the problem that had yet been made, and it is not important that it is not the final word. Two points of view are not contained in these distinctions: the primacy of the etymological principle over the two that we would today call the psychological and the physiological, and the presence of the physical point of view that we would today call acoustic. However, the realization that it is a question of several points of view here, which are by no means unrelated but whose "coincidence" is "to be considered as accident"—this could hardly have been more clearly and concisely expressed.

Some five months after this article, Latham's work *The English Language* appeared. There he writes:

> There is a difference between a connection in Phonetics and a connection in Grammar. Phonetics is a word expressive of the subject-matter of the present Chapter. The present Chapter determines (amongst other things) the systematic relation of Articulate Sounds. The word Phônaeticos (φωνητικός) signifies appertaining to Articulate Sounds. It is evident that between sounds like b and v, s and z, there is a connection in Phonetics. Now in the Grammar of languages there is often a change, or a Permutation of Letters: e.g., in the words tooth, teeth, the Vowel, in price, prize, the Consonant, is changed. Here there is a connection in Grammar... The Greek language changes p into f. Here the connection in Phonetics and the connection in language coincide. The Welsh language changes p into m. Here the connection in Phonetics and the connection in language do not coincide.[31]

1848 saw the publication of a book by A. J. Ellis on phonetic spelling, with the title: *A Plea for Phonetic Spelling* (2nd ed.). In the same year, his *Essentials of Phonetics* appeared. In 1849 *The Teacher's Guide to Phonetic Reading* (2nd ed.) was published. The *Phonographic Journal* was founded in 1842; from the 2nd volume onwards it was called *Phonotypic Journal*, then its title changed in the 9th volume (1850) to *Fonetic Journal* and in the 11th volume (1852) to *Phonetic Journal*, which title is retained until 1899.[32] In 1854 Ellis' *English Phonetics* appeared. In 1856, an essay appeared by Raumer, "Consequenzen der neuhistorischen Rechtschreibung und das historisch-phonetische Princip."[33] The concept of "phonetic" was used more and more in the 1850's for practical questions of writing in accordance with pronunciation.

In the following years the term acquired a new meaning. In 1856 and 1857, the two crucial speech-physiological works of Ernst Brücke[34] and C. L. Merkel[35] appeared, which, through their influence on W. Scherer, were influential also on the development of the linguistics of the "Junggrammatiker."

Merkel had reviewed Brücke's "Physiologie und Systematik der Sprachlaute" a year after its appearance in *Schmidt's Jahrbücher*.[36] The Viennese answered the Leipziger the same year with his essay,

"Phonetische Bemerkungen," [37] to which Merkel replied in the following year with a new essay, "Über einige phonetische Streitpunkte." [38]
What interests us here (the progress of the debate may be gleaned from
the original articles) is that phonetics is conclusively used as synonymous with the physiology of voice and speech—the way having been
prepared by Merkel's "Anthropophonik." Thus, the way in which
research was developing prepared the way for a second change in
meaning: Brücke already had made experiments with the kymograph
invented by C. Ludwig;[39] Rosapelly and Havet had undertaken experimental investigations of the activity of the larynx and lips in 1876;
Barlow had worked with the logograph in 1878. The physiological
concept of phonetics, quite of itself, became that of an experimental
method—that of experimental phonetics.[40]

"When Sweet laid the scientific foundation of English phonetics in
1877 and at the same time gave the concept of 'phonetics' general
currency, he utilized the preliminary studies of the Englishmen Ellis
and Bell, and the speech-sound physiologies of the Germans Brücke
and Sievers." [41] When in 1881 Sievers called the second edition of his
Grundzüge der Lautphysiologie "Phonetik," [42] word and concept
were naturalized in the entire scientific world.

A concise survey of the number of works which—according to my
calculations—bear the words "Phonetik," "phonetisch," "phonetic,"
"phonétique," etc., in title or subtitle, serves to illustrate this naturali-
zation:

1821–30	3	1861–70	10
1831–40	0	1871–80	20
1841–50	4	1881–90	76
1851–60	6	1891–1900	53

A second survey, of the first appearance of this word in scientific
works in the various countries (similarly restricted to works having
the word in title or subtitle), will illustrate the rapid spread of this concept throughout the scientific world:

1822	France	
1825	England	
1856	Austria	
1856	Sweden	
1859	Germany	
1877	U.S.A.	
1884	Portugal	
1890	Switzerland	
1892	Chile	
1894	Spain	
1897	Denmark	

2. ANCIENT PHONETICS

The well developed phonetic systematization of Vedic antiquity only became accessible to phonetics through the editing and translating of the *Rig-Veda-Prātisākhya* by Max Müller,[1] even though Ernst Brücke[2] had studied the phonological system of Sanskrit earlier (in 1856) with the help of the investigations of Bopp, Benfey, and Böhtlingk and earlier works of Müller. The spiritual power of the word in connection with the mythical and magical *weltbild* of the Vedic religion has been treated and described by Ernst Cassirer.[3]

Remarks on speech and singing techniques and on phonetic views among the Greeks and Romans are to be found above all in Lersch,[4] Volkmann,[5] Steinthal,[6] and Stadelmann;[7] and in connected form in the work of Krumbacher.[8, 8a]

The Greek conception of the meaning and function of language—especially that of Plato and Aristotle, who did not develop an independent linguistic philosophy of their own—is described by Ernst Cassirer[9] and Julius Stenzel,[10] to whose investigations we shall refer below. Particularly in Stenzel, considerable space is devoted to the Greek thinkers, in his treatment of questions of contemporary linguistic philosophy, since for problems of the philosophy of language no philosophy provides a firmer foundation than one in which the separation of linguistic thinking from the total sphere of cognizance has not yet taken place; one which instead in its concept of the Logos keeps continuously in mind the verbal side of each genuine philosophical thought; one which never treats the problem of truth independently of the word which signifies and contains truth, and for this very reason would not consider treating the problem of language, of the communicating word, independently of the truth apprehended in the word and only thus comprehensible; nor independently of the object to which speech refers, and of the discourse in which by speech and counter-speech, by reasons and counter-reasons, the subject in question gradually stands out more clearly.[11]

In the context of a history of phonetics, the philosophy of Heraclitus is of decisive significance. At first sight, the concept of the "Logos" appears to be still closely related to the mythical view of the dignity and omnipotence of the heavenly word. However, in the midst of the language of myth which Heraclitus still speaks, a new tone is perceptible. In full consciousness and clarity for the first time, the philosophical and speculative fundamental idea of the uniform and inviolable regularity of the universe is contrasted with the mythical view of the cosmic process. And this belongs to the tasks of phonetics or phonometrics, respectively: to see and grasp the relationship of word and language to the natural (i.e., in accordance with the laws of nature) determination of the production and transmission of sound—at the level of research, naturally. In the Heraclitean idea of the Logos, language and regularity combine for the first time—even if initially from quite a different viewpoint. But, however far he may be from the idea of research, his idea of "invisible harmony" is close to it. "As Heraclitus puts the single object in the continuous stream of development, and allows it to be destroyed and preserved in it at the same time, so the single word is assumed to stand in relation to the entity of 'speech.' " [12] Plato's relations to speech and the concept of speech have been made the basis of linguistic philosophy, especially since Julius Stenzel.

How close the analysis of the logical forms is to that of the linguistic ones in Aristotle is indicated by the designation of the "categories," which represent the most general relationships of being, "which as such imply at the same time the highest types of assertion (γενή or σήματα τῆς κατηγορίας);thus in fact the construction of the sentence and its division into word-units and word-classes seem to have been in many ways a model for Aristotle in the formation of his system of categories. . . . Logical and grammatical speculation therefore seemed to correspond completely here, and to condition each other reciprocally—as the Middle Ages, following Aristotle, clung to this correspondence." [13]

In Plato and Aristotle, the art of grammar (τέχνη γραμματική) comprised the whole academic consideration of sounds, including the physiological side of speech and theory of accents, and was connected with metrics and music. Indeed, the more precise or real phonology was a part of metrics, and the inventors of phonology were metricians.

One part of the recognition of melody (ἀνάγνωσις κατὰ προσῳδίαν) was the theory of accents: it trained hearing and speech; it also taught one to grasp and to master the finer distinctions of the speech-melody. The designations of the accents, however, were adopted from musical technique and gave expression to the distinctions "higher and lower accentuation of individual syllables." [14] Thus, for example, the musical terms ὀξύ, high, and βαρύ, low (in "Philebus") were transferred by Plato in the "Cratylus" to the word-accent: high tone (ὀξεῖα), low tone (βαρεία).

However, a real theory of acoustics or waves was not developed in antiquity at all. There are only Pythagorean theories on the acoustics

of music, for—according to the investigations of E. Frank[15]—the theory
of proportion, which dominated the whole mathematical thinking of
the Greeks, developed among Pythagoras and his followers on the basis
of music. With regard to acoustics, Plato in the *Republic* also follows
the Pythagoreans. Moreover, in the "Timaeus" he has left us a table of
tones very close to present-day conceptions: rapid vibration produces
a high tone, slow vibration a low one; the tone is transmitted by vibrations of the air from the body causing vibration to the ear, from whence
it is transmitted to the brain.

Whereas Anaxagoras[16] imagined that the production of voice is
caused by a collision of breath (πνεόματος) against the stable air and by
an echo-like return of the shock to the ear, Plato[17] defines φωνή as a
shock to the brain and blood brought about by air and communicated
through the ears, which then transmits itself to the soul. A very clear
distinction was, however, made between speech, voice, and sound. One
is reminded of Latham, when one reads, for example, from Aristotle:
"φωνή καὶ ψόφος ἕτερον ἑτέρου ἐστι καί τρίτου τούτων διάλεκτος" (Speech and
noise are different from one another, and language differs from these—
Historia animalium, 535ᵃ, 28).

The intervals of the keys were given by the Greeks in ratios (2:1:3/2
etc.) that also represented the numbers of vibrations.

The difference between the melody of the spoken word and that of
the sung word was defined by Aristoxenus in his first Harmonics: Both
voices perform a topical movement (that is, proceeding by height and
depth)—the speaking voice a continuous movement, the singing voice
a discontinuous one. In speaking, therefore, the voice does not remain
on one level, but is in motion until it ceases; in singing, it dwells on
certain tones.[18] And concerning the compass of the singing voice,
one reads in Dionysios: "The interval by which the speech-melody
[διαλέκτου μέλος] rises and falls is closest to a fifth: It does not rise more
than three whole-tones and one half-tone, nor sink more than this interval. The syllables of a word are not on the same level, but some on a
high level, others on a low level, a third on both at the same time." [19]

The views of Democritus, Epicurus, and the Stoics, and especially
the essentially more fact-based theories of Aristotle have been described
in detail by Lersch.[20] It is significant for the connection with the modern development of physical acoustics that the Greeks began with the
tetrachord and combined two tetrachords into a *scale*; for example,
into the Dorian *e f g a / h c' d' e'*.[21] Since they established no absolute
pitch, this scale can be formed, beginning on any base-note, in the

order: whole tone, half tone, whole tone, half tone, whole tone, whole tone. A more detailed exposition of this belongs to the history of music. In addition, it is significant "that the ratios of length are established in the monochord, and the numbers of vibrations are recognized in inverse proportion to the length. Thus, although vibrating strings were of fundamental importance as sound-producers for mathematical developments among the Greeks, we have no theory of the process of vibration preserved in the literature which remains to us." [22]

One may say that such opinions persisted until the acoustic and wave theories were taken up in the circle of Galilean research. From the answer that Aristotle gave to the question of why the waves sometimes arrive sooner than the wind,[23] it is clear only that he thought the waves continue the impetus that caused them; for the rest, however, his explanation is unintelligible in terms of physics.

3. RENAISSANCE PHONETICS

Here we need merely mention the presentations of purely linguistic views and endeavors in medieval and modern times, of which presentations there is no lack: to Benfey,[1] Scherer,[2] Raumer,[3] Delbrück,[4] Hermann Paul,[5] Konrad Burdach.[6] However, in general, all these works go into phonetic investigations (in the strict sense) only after the period when Scherer had begun to make Ernst Brücke's works on the physiology of sounds (in connection with endeavors to ascertain the phonetic value of letters and to seek reasons for the sound-shifts) profitable for linguistics. And the extensive physical works on the movements of air molecules to be associated with sounds began to be included, at least peripherally, in the circle of linguistic deliberations only after Sievers[7] and Rousselot.[8] But all these physiological, anatomical, and physical investigations in turn have a long previous history. It is all the more astonishing that linguistics paid them such little attention since a natural conception of language—if often only in the form of vague images and analogies—had already played such a decisive role in the Romantic beginnings of linguistics. And even if it is true that, at the beginning of the 19th century at least, some linguists still had just as little academic contact with contemporary physicists, physiologists, and mathematicians as the latter had with them, the basic intellectual attitude of the epoch binds all the sciences together so firmly that at least a retrospective examination should not leave so neglected these endeavors to investigate speech and speech-organs, which lie outside linguistic science in the narrower sense. The history of linguistics itself gives decisive indication as to the direction in which the contemporary relation of these various scientific endeavors is to be sought. Just as the linguistic and philological endeavors of the 16th and 17th,[9] 18th, and 19th centuries can nowhere be completely separated from the basic philosophical attitude of the time, so also is prevailing physiological and especially physical research indebted to the philosophy of its time. This is true not only of Humboldt's influence on Bopp, Pott, and Curtius, or of the dependence of the young Schleicher on Hegel; this is true of all relations between proposition and decision since the earliest linguistic deliberations of the Greeks. However, there is unfortunately no comprehensive account of a history of modern philosophy of language. Even Ernst Cassirer in his treatment of "the language problem in the history of philosophy" claims "only to pick out the main impulses in the philosophical development of the 'language idea' and to lay down some preliminary guiding principles for a future detailed treatment of the subject." [10] And even if Cassirer's concept of language (which is close to Humboldt's) is not shared by us, because we cannot grasp the concept of language without that of linguistics, Cassirer's presentation of the history of linguistic philosophy shows most clearly, as does his whole work, the close connections of the individual sciences to one another.

The following remarks are therefore restricted essentially to particular sides of this question that have not been more closely considered, either in Cassirer's work, or in the other investigations of the philosophy of language, or in the well-known presentations of the history of linguistics.

When, at the beginning of the modern period, the battle against Aristotelian logic began, when its right to be called *the* systematization of the mind was challenged, the close alliance it had entered with language and general grammar formed one of the most important and most dangerous points of attack. "On this basis Lorenzo Valla in Italy, Ludovico Vives in Spain, and Petrus Ramus in France attempted to revolutionize Scholastic-Aristotelean philosophy. At first the battle was fought within the limits of philological research and views of language: it is that very 'Philology' of the Renaissance which, from its deepened view of language, demanded a new 'Epistemology.' What Scholasticism had comprehended of language, as is now objected, was only the external grammatical relationships, while the real core, which is to be sought in stylistics rather than in grammar, remained hidden from it." [11]

However, these endeavors led to the idea of research in the natural sciences, as in modern physics and comparative biology, and in humanities, which developed into history, literature, and linguistics. This was made possible by the shaking of the Neoplatonic-Christian edifice of ideas by Nicolaus Cusanus. The fact that religious perception has its own roots in quite different realms of the mind from natural intellect and logic would have been gathered directly from the Pauline writings,[12] but it was obscured by the uniform *weltauffassung* of Neoplatonic-based theology. The path for these thoughts of Cusanus was paved in Occam, as Gerhard Ritter's[13] investigations of the struggle of the *via antiqua* with the *via moderna* in the 14th and 15th centuries have shown.

And as through Cusanus the monodimensionality of medieval universal knowledge was ended and logic removed from theological and metaphysical speculation, so through Galileo physics was torn away from logic and philosophy—or rather, it was placed in a defined relationship to them.

Considering the decisive importance which the concept of the physical experiment has for acoustics on the one hand, and for a critique of "experimental" phonetics and its nature-based conception of language on the other, it is necessary to consider in rather more detail the achievement of Galileo. This does not consist solely in the discovery of the first empirical scientific method[14] in the modern sense; through it the procedure of each individual inquiry has entered into a new relationship with philosophy, which has its roots in Plato's "Meno" and reached its culmination through Kant's investigation of the presuppositions of Newtonian physics.

Comparative linguistics is also an empirical scientific method—although of a different tendency and kind. The fact that it could discover its real task in Bopp, Rask, and Grimm at the beginning of the last century may be attributed in no small measure to Galileo, who discovered such a goal for the first time and extricated it from philosophical speculation.

Phonometrics can fulfill its aim of investigating the natural side of speaking only if it presupposes the idea of language and linguistic intelligibility, then submits to the thought of linguistics, and thus in the final analysis attempts to practice not physics, but linguistics. To be physics, to be an exact scientific investigation of Nature—perhaps also, by the exact analysis of the vibrations of air molecules, to replace or dispossess linguistics—that has been the hope and goal of experimental phonetics in the past and is partly so even today. In its very name and, as will be shown, in many of the questions it poses, this wish is expressed. It is to be shown in this work that this desire has no scientific goal. One of the ways to do this will be to show that in the recording methods of experimental phonetics it is never a matter of experiments in the physical sense of the word. For this reason it shall be explained here what has constituted the concept of the physical experiment since Galileo.

We cite below the decisive paragraphs on this topic from Hönigs-wald's investigations on the history of the problems:[15]

With Galileo, the question of method enters into a direct relation with the tasks of scientific research: to seek laws of nature already presupposes criteria for their universal validity.—Now it is a matter of removing the phenomena from the entanglement of chance circumstances—i.e., to define them, to reduce them to their concept. But how is this goal to be attained? Certainly not by collating and comparing many cases of the phenomenon in question—as is well known, it is the question of the free fall of bodies—and by examining them with respect to their common characteristics. For the fact that one characteristic is common to many cases does not by any means establish the regularity of a process, although certainly all the cases will satisfy regularity. The law will consequently have to be common to all the cases. Only the law can justifiably claim to be valid in every possible case but not those characteristics that—perhaps only by accident—are common to all cases observed. Galileo speaks in a completely new sense of cognition from experience (or, if one prefers this term, of induction), in a sense which now guarantees a quite new meaning to the result of induction also. For the law towards which he advances, or wishes to advance, is not a mere uniformity of events. It is much more; it represents insight into the conditions for uniformity. However, for this very reason it means, according to the idea of the law, the concept of the phenomena which satisfy its condition; and a theory of the law of nature is seen as part of a theory of the concept, i.e., as logic. Not only does the indissoluble relationship between logic and the theory of method appear established by this for all time; above all, the path of research itself is sketched out in the most remarkable way. "Law" is now equivalent to the sum total of stipulations which condition the phenomenon as the very phenomenon which it is. The law of the free fall of a body is seen as the condition according to which a natural event is first characterized unequivocally (i.e., objectively) as "free fall of a body." Not only, therefore, must each individual case of a phenomenon underlie the natural law; the law must also be demonstrable in each individual case. Galileo had thus to solve an absolutely new methodological problem: the logical analysis of the individual case. If it is completed, then the law is found which reaches beyond all the empirical differences of the individual cases. For this very reason, each attempt of Galileo to advance towards the law of the free fall of bodies appears at the same time as an attempt at such an analysis.* In this sense Galileo is not concerned with the knowledge of the process of many falls, but with the cognition of one fall. For it entails the cognition of all cases of a phenomenon. Moreover, induction—understood as the subtraction of the individual case—appears impossible or pointless. It is impossible if the number of cases in question is infinite, useless if this number is limited. For in the first case the process could never be completed, in the second the result would already be presupposed in the preliminary propositions.** Galileo certainly investigated many cases. But the bias, the meaning of such research passes on to the analysis of the individual case. Many cases are investigated, but only to ascertain the structure of the individual case. By the restriction on this task Galileo excludes "experience" no less. On the contrary: he makes it the source of strict scientific cognition only in extricating the individual case from the entanglement of chance circumstances. All his methodical measures have the same aim: to comprehend the individual case in its conceptual purity i.e. in its necessity. For this reason, however, experience is a systematic unity, a methodical connection of analyzed material—not an unplanned heaping up of unanalyzed and logically separated material. Experience too means precisely cognition; experience too is characterized by necessity, although in a particular way. Thus the exclusion of chance circumstances, which "disturb" the clarity of the cognition of the law, is an essential element of his whole methodical procedure.

In no method can the experiment play a greater role than in Galileo's. It turns out to be the decisive and last court of appeal in the exploration of the laws of nature. However, in spite of, or even because of this, it only reflects the basic intention of the Galilean methodology. Here one does not make experiments in order to prove the regularity of the phenomenon in question by a comparison of their results, but they are seen as representatives of the very regularity of the phenomenon itself. In other words, they are not a foundation on which cognition of the law is based, as with the traditional induction; they are rather the empirically palpable expression of that cognition. They do not establish the law, but they "verify" it, or they establish it only insofar as they verify it. The experiment repeats the conditions which the law expresses "hypothetically" with regard to the experiment. The individual phases of the investigation—the hypothetical statement of natural regularity, the development of its consequences, and the verification in the experiment—are only temporally separated from each other. Essentially they belong together and are inseparable. Because and insofar as the individual case—at least in theory—exists in its own nature, separated from all chance (i.e.,

---

* Cf. on this Paul Natorp, "Galilei als Philosoph," *Philosoph. Monatsh.* 1882; Alois Riehl, Über den Begriff der Wissenschaft bei Galilei. Vierteljahrsschr. für wissenschaftl. Philosophie; further, my work: Beiträge zur Erkenntnistheorie und Methodenlehre 1906.

** Galileo, Versus Vincenzo di Grazia, Op. XII, 513.

in its conceptual determination, its "necessity"), then necessity is inherent also in the deduction from the one analyzed case to all cases. Completely new concepts of "induction" and of "experiment" are thus introduced. The former is no longer the comparative counting up of cases, with the aim of "abstracting" the law from them; and the latter has ceased to be nothing more than "a clever question to Nature." Induction—i.e., the investigation of laws from experience—appears completely fused with the "deduction" of the consequences from the first proposition of the connection to be verified in the experiment. Induction has changed into the "inductive-deductive" process, as it has been called. Or in other words: by reason of the naive experience that bodies deprived of support fall to the ground, a certain and well defined relationship is credited to the phenomenon as a law, with the methodical intention of eliminating chance circumstances. One therefore "ascends" from the unanalyzed cause of the investigation to the necessary law and "makes the induction"—*within the thought*—to descend at once to the experiment, i.e., to the experience analyzed in the sense of a hypothesis. However, ascent and descent are only tendencies of a state of affairs which can only authenticate its methodological sense as a whole. The word "deduction" must therefore also undergo a fundamental change in meaning. Deduction which would be capable of maturing new insights is analysis. It is the method of examining a given fact in respect of its conditions, of comprehending it from its conditions; i.e., in its conditionality. Mathematics has become the model for deduction; and the great thing about Galileo's method is this, that its originator knew how to permeate experiment with the original mathematical motive of analysis and thereby to give its concept a new methodological content. For this very reason, however, experiment changed into "clear-sighted intervention," by which "simple forms of events are isolated, in order to submit them to measurement." *

But the concept of the hypothesis also takes on a new and significant meaning at the same time; or rather, it returns to its oldest and essentially original meaning—the Platonic meaning. For "hypothesis" means within the context of the Galilean method, just as in Plato, "laying the foundation." A proposition is seen as "hypothetical" not because it is unfounded, but because it lays claim to being proof itself. Certainly, this claim can only be regarded as fulfilled if it is shown in experiment to be authentic. However, as surely as the experiment in the context of the Galilean method means only another expression of the conditions, that is, of the concept of the phenomenon in question, just as surely is the methodological sense of that claim established independently of the authentication of the hypothetical proposition by the experiment. The proposition, that the velocity of a free-falling body increases with the time of fall, is a "hypothesis" insofar as it contains the claim to define the phenomenon of the free fall of bodies—i.e., to define them free from all chance and disturbing circumstances. Only with respect to this claim then can one

* Windelband, Geschichte der Philosophie, 2nd ed. 1900, p. 131.

speak of "verification" of its consequences by experiment—as with the help of Atwood's fall machine. Experiment and hypothesis are reciprocally related. If they contradict each other, then their concepts themselves are impossible; i.e., the conditions of the procedure are unfulfilled, and the investigation must be begun all over again. In a certain phase of Galileo's investigations this was, in fact, the case. For his first assumption, that is, that the velocity of free-falling bodies increases with the space traversed, no experimental confirmation was offering; it then ceased to be a "hypothesis." "Hypothesis" requires an experiment; i.e., it lays down its conditions. But at the same time it stipulates its technical assumptions. For instruments are as little part of or external to it, as the results of research. It is rather a product of the method, the expression of the intellectual process itself become visible or "material." * In the concept of natural law "thinking" and "experience" for Galileo combine in an indissoluble connection. In this concept he is conditioned also by his relation to traditional logic: the consciousness of the conditions of their objective definition must give way to the bloodless "abstraction" from things. This however comprises the full reality divested of all contingencies. The stricter and more consequential the analysis, the more sharply defined by this is that reality. Logic is itself possible only with respect to the conditions of cognition, it has really become now for him the " 'organon' of discoveries." **

However, the new concept of the perception of nature demands a new concept of nature. How is the concept of nature defined, so far as the perception of nature is accomplished according to Galileo's method? One sees that the problem of the exact sciences is at the same time the problem of their possible objects. The unity of cognition conditions the unity of nature. The concepts, in which the latter is seen, only fulfil the requirements of scientific cognition. This is the way in which the much discussed "materialism" of Galileo, as well as his atomic theory, should be judged. Matter and atoms are not existential values independent of all perception. Their concepts are seen rather as means for the ordering and cognitive subjugation of multiplicity, as bearers of the thought of necessity.*** They cannot be assessed differently from Galileo's theory of the subjectivity of sensual qualities: the unambiguous nature of cognition requires that in investigating the regularity of nature, the sensual determination of colour and tone is disregarded, and only the "necessary" state of things, as they are seen from the point of view of number, time and space, is to be taken into account. And exactly the same applies to motion and inertia. They too are not given states of some kind or other. They too are only valid so far as mathematical requirements are satisfied in them, so far as they can be demonstrated in their necessity and unambivalence. The "analytical" method also proves to be authentic: mo-

---

* Alois Riehl, Philosophischer Kritizismus, Vol. II, 2 Leipzig 1887, p. 4.
** Cassirer, op.cit. (Das Erkenntnisproblem in der Philosophie und Wissenschaft der neueren Zeit. Bd.I), S. 323
*** Dialogo dei massimi sistemi. Fourth day, Op. I, 497.

tion and inertia confirm experience because, framed in the mind, they bring the phenomena to objective determination. Everything which is opposed to this state of affairs is foreign to the scientific concept of motion. It is his task to comprehend not the undefined 'being' of motion, but its regularity only. Galileo's relation to metaphysics seems thereby to be called in question. His concept of a cognition of nature is the concept of a cognition of the order of phenomena. If this view is valid, then nature is to be defined as the object of possible cognition, and it is not the mysterious working of metaphysical powers. However, just as little does it allow exception to its laws. Nature itself means "order." Therefore the real law of nature "explains" exceptions no less than rules. For an exception is, like the rule whose concept it necessitates and supplements, a completely indefinite image so long as it is not comprehended from a law. Hence, every type of empiricism is for Galileo surmounted in principle. For it is precisely empiricism which, at least in the sphere of the cognition of nature, provides the strongest support for metaphysics. The empiricist appeals constantly to "facts" and thereby to the objectivity of nature. However, he has no possibility of even discussing the concept of fact, much less of exposing it in its strictness and purity. Thus in the final analysis he requires what he is ostensibly opposing: the assumption of a metaphysical "essence" of nature behind the regularity of events, in order then to renounce cognition of this being immediately, whether joyfully or resignedly. Galileo is quite different. In the investigation of the laws of actuality, the "fact" itself becomes a problem for him. He comprehends its concept in that of research and thus gives a new and critical content to the old Parmenidean thought: τὸ γὰρ αὐτὸ νοεῖν ἐστί τε καὶ εἶναι [for it is the same thing to think and to be]. According to his concept, objective existence is in accordance with the conditions of cognition.[16]

H. Arens, in his history of the problems of linguistics which is distinguished essentially from others in the breadth of the disciplines included and to that extent contains much important material for our consideration in this chapter, points out that already in the 16th century—that is, in the time of Galileo and his discovery of the so-called Galilean laws—the Florentine merchant Filippo Sassetti refers for the first time to the connections between Italian and Sanskrit in a letter from Portuguese India. In this letter Sassetti says: "Our present-day language has much in common with it, among other things many words, especially numerals: 6, 7, 8 and 9, *Dio*, *serpe*, and many others." [17] It is, of course, not easy to ascertain today what Filippo Sassetti meant by the words "among other things" and "much" in this passage of his letter; that is, whether he noticed other "similarities" besides the "many words." In any case, this letter not only seems to contain the earliest reference to the relationship of Sanskrit with a European language, but also to have expressed this discovery, of necessity, in a simple statistical form not yet noticed by linguistics. For there is no doubt that a statistical problem is concealed behind his assertion of "many" words in common, especially since Sassetti on the basis of his observation asserts common features of the two *languages*, not of counted Italian and Sanskrit *texts*; that is, he relates frequencies which he can only have obtained by counting or estimating on a limited number of texts, to the two languages themselves—i.e., to frequencies in *not* counted or estimated texts. It is a statistical problem, not noticed by linguistics to be such until recent times, that is concealed behind such a deduction, without which linguistics is not "possible" as a science.[18]

4. PHONETICS OF THE SEVENTEENTH CENTURY

Descartes—like the Greeks—did not make language the subject of independent philosophical investigations. However, in a letter to Mersenne, which Cassirer mentions, he places the problem in a decisive context: "To the demand for 'mathesis universalis' is added the demand for a 'lingua universalis' . . . As a quite precise order exists between the ideas of mathematics, e.g., between the numbers, so the totality of the human consciousness with all the contents, which can ever enter the same, forms a strictly ordered total. As consequently the whole system of arithmetic can be built up from relatively few numerical symbols, so also by a limited number of verbal symbols, if these could only be combined in accordance with definite universally valid rules, it should be possible for the totality of intellectual content and its structure to be described exhaustively." [1]

The demands made here have their roots in the *lingua adamica* of Jacob Böhme. The idea of a universal alphabet had after-effects via Leibniz even in the phonetics of the 19th century and the present day—particularly in the ever-recurring attempts to put the sounds of all languages into a mathematical relation to each other in the form of vowel-triangles or vowel-polygons.

The first vowel-system appears to derive from Robinson in the year 1617, in contrast to which the vowel-rectangle of Wallis of 1653 appears to be an independent creation.[2]

The age immediately following Descartes did not let itself be disconcerted by the critical caution which finds expression in the Descartes writings.

In rapid sequence Descartes' successors produced the most diverse systems of artificial universal language, which, though very different in execution, were in agreement in their fundamental idea and the principle of their structure. They all started from the notion that there is a limited number of concepts, that each of these concepts stands to the others in a very definite factual relation of coordination, superordination or subordination, and that a truly perfect language must strive to express this natural hierarchy of concepts adequately in a system of signs. Starting from this premise, Delgarno for example, in his *Ars signorum*, classified all concepts under seventeen supreme generic concepts, each of which is designated by a specific letter; all the words falling under the category in question begin with this letter; similarly, the subclassifications distinguished within the common genus are each represented by a special letter or syllable affixed to the first letter. Wilkins, who strove to complete and perfect this system, established forty principal concepts in place of the original seventeen and expressed each of them by a special syllable, consisting of a consonant and a vowel.* All these systems pass rather hastily over the difficulty of discovering the "natural" order of fundamental concepts and of clearly and exhaustively determining their mutual relations. More and more their authors transformed the *methodic* problem of classifying concepts into a purely *technical* problem; they were satisfied to work with any purely conventional classification of concepts as a basis and, by progressive differentiation, make it serve for the expression of the concrete cognitive and perceptual contents.[3]

These ideas were still operative—although there may be no historical continuity involved—in F. K. Fulda's[4] attempts[5] to arrange all human concepts into a system.[6]

Having indicated here the *philosophical* roots that nourished the natural systematization of the speech-sounds of all imaginable languages, which flourished in the 19th century, we should now mention one of the chief sources of the physiology of speech—the teaching of the deaf and dumb, as founder of which the Benedictine monk Pietro Ponce may be regarded (Ponce died at Oña in 1584).[7] He is said to have written a book on his method, but it is not extant.

The oldest work on the teaching of deaf mutes is by Juan Pablo Bonet,[8] who deals with the Spanish speech-sounds, their symbols, and a phonetic reading method recommended by him, and who further— in connection with advice on the teaching of speech—gives a physio-

---

* If the letter P, for example, signifies the general category of 'quality', then the concepts of size in general, of space and of weight are expressed by Pe, Pi, Po, and so on. Cf. George Delgarno, Ars Signorum vulgo Character universalis et lingua philosophica, London 1661; and Wilkins, An Essay towards a Real Character and a Philosophical Language, London 1668. A short outline of the systems of Delgarno and Wilkins is given in Couturat, La logique de Leibniz, Paris 1901, Note III and IV, p. 544 ff.

logical phonology which describes the positioning of the vocal organs for the individual lettters treated by him.

In England, independently of these Spaniards, physiological phonology and its practical application was founded by Bishop John Wallis, who preceded his *English Grammar* (which appeared in 1653) by a "Tractus grammatico-physicus de loquela." [9] If we exclude the purely physiological investigations in the 17th Century, we may describe this book as one of the earliest phonetic works in the stricter sense.

Franziscus Mercurius van Helmont carried the natural-philosophical and theosophical doctrines of his father Johann Baptist van Helmont further and, already before Leibniz, had developed a monad theory expressed in his investigations of the philosophy of language. This friend of Leibniz, to whom the latter dedicated an epitaph, was born at Vilvorde near Brussels in 1614 and was introduced to medicine and chemistry while still at home. He devoted himself later to the study of theodicy, and in 1667 published his draft of the "Original Alphabet of the Holy Language." [10] For him, too, the "original" language was Hebrew, the language in which, according to the prevailing view, God had spoken in Paradise to the first human beings. This "original" quality was also claimed for Hebrew script by the writer, following in the footsteps of his father, the mystical theosophist and doctor, Johann Baptist van Helmont. For him the Hebrew letters possess, at least in a prototype to be discovered, the value of articulated script.

Through Leibniz, two new viewpoints then appeared in the shallow pseudo-scientific activity which—following Descartes' lead—had developed into the construction of artificial universal languages: Leibniz posed the linguistic problem again, as Descartes had done and as the Greeks had done, in the context of all theoretical cognition, of the logic and idea of science; in this context again, he brings to the concept of language a quantitative tendency from a newly gained depth: the newly established mathematical analysis, which is distinguished from all earlier numerical speculation by its belonging—after the establishment of the physical method by Galileo—to the earliest products of Western research that are science in the modern sense of the word. Here a foundation has been acquired that indeed does not yet allow (as he thought) the broaching of the problem of language. It was not yet the method of linguistic research, but it was a truly scientific, indeed a mathematical method, and thereby it became one of the firmest foundations of every later form of quantitative treatment more adapted to the subject—including the treatment of speech and speaking.

And yet another uncommonly important perception meets us in Leibniz for the first time: he does concede to Descartes that the genuine universal language of cognition is dependent on cognition, that is, on "true philosophy," but he adds that it nonetheless need not wait for the perfection of this; rather, both actions, the analysis of ideas and the allotting of symbols, can develop in and with each other. Here Leibniz removes the problem of language—if only in the philosophical-speculative form of his time—from the stricter problem of philosophy. Nothing was more obvious to Leibniz than subsequently to anchor the idea of language to a methodical certainty that was already convincingly authenticated in his system of the intellectual world: to the idea of number and mathematics. "Vetus verbum est, Deum omnia pondere, mensura, numero fecisse. Sunt autem quae ponderari non possunt, scilicet quae vim ac potentiam nullam habent; sunt etiam quae carent partibus ac proinde mensuram non recipiunt. Sed nihil est quod numerum non patiatur. Itaque numerus quasi figura metaphysica est, et Arithmetica est quaedam Statica Universi, qua rerum potentiae explorantur." [11] Certainly, such reflections cannot serve us as a guide for our present-day scientific work: they impart to us neither the cognition of actualities nor the method or technique by which to master them. But it is clear what is involved here (in this form and resolve for the first time): it is the idea of an exhaustive intellectual mastery—and by this Leibniz means a mathematical mastery—of the final order, which must be at the same time the order of being, of thought, and of language. It is still a long way from the perception that there is no absolute control of objective actualities, that the idea of their mastery is nothing more than the idea of endless research (i.e., of endless proof), and that the orders of being, thought, and language—although in no way independent of each other—are yet distinct from each other (namely, that they are scientific tasks distinct from each other, and are themselves subject to *a final* order: the order of the sciences). Physics is one of these tasks; mathematics—not independent, but distinct from it—another. These two were known to Leibniz. The next to be recognized were chemistry[12] and comparative biology, and only around the turn of the 18th century did history and philology join them as independent tasks; and psychology followed them with the definition of its own task in our times.

We should at least mention the work of Schelhamer of 1677,[13] and that of Narcissus, Bishop of Ferns and Leighlin, of 1684.[14]

The first to go beyond the experiences of the Greeks in relation to

the monochord was Bacon of Verulam; certainly, he too investigated
these oscillations essentially through the tones produced. He thereby
established the influence of tension, "temper," length, and diameter.
In 1635 Mersenne undertook measuring experiments, among them the
two propositions: that the frequencies of strings of equal length and
tension are related as the square roots of their own weights, and that
with strings of equal length and thickness the frequencies are related
as the square roots of the tensing weights. He also noticed—without,
however, giving the reason—that besides emitting the key-note, an oscillating string emits the twelfth and still higher tones.

Newton appears to have been the first to occupy himself experimentally with the formation of water-waves.[15] He begins with the
propagation of a shock in a fluid. A diagram drawn by him makes clear
his theoretical view of this propagation. Newton defines the wave-
length as the distance between two wave-peaks or troughs. His view
is only approximately correct—as he himself says—since he assumed a
purely vertical oscillation for each water-particle, whereas in reality
it would have to vibrate in a circle. The theory was developed further
by Gravesande and later by Bernoulli.

The development of these theories up to the end of the 19th century
has been treated by E. Hoppe in his history of physics,[16] on which my
last remarks are based. In the following, we can only indicate and em-
phasize again how little notice of these investigations was taken by
linguists and by teachers of music and speech. Certainly, writers on
music such as Zarlino (1599) and Werkmeister (1691) treated singing,
but no physical investigations were made in conjunction. The theory
of speech-melody also remained the subject of speculation in the service of speech-education and elocution: Sheridan[17] (who in 1762 distinguished between accent, which unites syllables into words, and
emphasis, which unites words into sentences) and Schocher[18] point out
in 1792 the usefulness of reproducing spoken sentences by notes, and
the author of *Grundriss der körperlichen Beredtsamkeit*[19] gives examples of such notations in 1792. Hänle[20] too, in 1815, gives an appendix
set in notes, in which he cites rest, situations, moods, and questions of
narration and of memory, but, as Merkel[21] explains, "with a certain
musical prejudice, which often causes him to overlook or to forget the
real rhetorical modulation." Similar notations follow then in Michaelis
(1818),[22] Thürnagel (1825),[23] and Diesterweg (1830).[24]

5. PHONETICS OF THE EIGHTEENTH CENTURY

The "Dissertatio de loquela" by Johann Conrad Amman,[1] dating from the year 1700, is of particular interest from the point of view of the history of problems, because here for the first time, as far as we know—that is, about 250 years ago—the thought of breadth of variation in the pronunciation of individual sounds is at least hinted at.

We quote the beginning of the decisive second chapter, first in the original Latin text, then in English translation, arranged by Charles Baker, of the year 1873:

Caput II. Literarum naturam, et varioseas formandi modos exponit.— Hactenus de Loquelae subjecto generaliori, Voce nimirum et Spiritu, et utriusque formandi ratione, et scitu summopere necessaria differentia: inquirendum jam, quo pacto dicta Vox et Spiritus, ceu apta materia, in has illasve literas fingantur; Unica enim literarum materia Vox est et Spiritus, forma autem earum, ab organorum et meatuum, per quos transeunt, diversa configuratione petenda est. Literae igitur, non qua sunt Characteres quidam calamo picti, sed quatenus enunciantur, sunt Vox vel Spiritus, vel utrumque simul, diversimode ab organis Loquelae destinatis configurata.—Literarum possibilium numerus vix definiri potest, tot enim esse possent, quot diversis modis Vox et Spiritus ab organis Loquelae figurari queunt: Praeterea plures literae suam habent *latitudinem*,[2] et veluti gradus quosdam, idemque Character, etiam in eadem Lingua, non una eademque oris configuratione pronunciatur, sic a et e aperta quandaque audiunt, quandoque clausa, habet et o cum i suam latitudinem etc. Haec autem differentia, si variae Linguae invicem comparentur, majoris longe est momenti, et praecipua causa, quod peregrinas Linguas tam difficulter pronunciare discamus, antequam organa nostra ad insolitum aliquem motum flectere possumus.

Hitherto our attention has been directed to the general subject of speech, viz., voice and breath, and the method of managing both, and especially to the difference between them, which it is necessary to know. It remains now to be investigated by what contrivance voice and breath, as apt material, may be formed into the several letters; for, the sole elements of letters are voice and breath, but the form of them is to be ascertained from the different configuration of the organs and passages through which they are transmitted. Letters therefore, not characters, which are written as with a pen, but as enunciated, are the voice and breath, or both together, differently modulated by the requisite organs of speech.—The number of possible letters is scarcely able to be defined. It may extend as far as the voice and breath may be varied by the organs of speech. Besides, many letters have their own *compass*[3] and peculiar modifications; and the same character, even in the same language, is not always pronounced in one and the same manner. Thus at one time a and e are open, at another shut; o and i have their own modifications, etc. But this difference, if several languages are compared together, is of far greater moment, and is the chief cause why we learn with so much difficulty how to pronounce foreign tongues, before we have adapted our organs to some unaccustomed movement.

Wilhelm Viëtor, who arranged for the reprints of the Latin original and of Venzky's German translation[4] of 1747, wrote already in 1917 in the introduction to these reprints, that one is "really astonished to find, e.g., Sievers' 'gewisse Breite der Richtigkeit'[5] already in Amman at the beginning of the second chapter." Today one would translate "latitudo literarum" most succinctly by "breadth of variation of the realization of sounds."

Then one should mention the works of Dodart,[6] Lischwitz,[7] Ferrein,[8] Runge,[9] de Brosches,[10] Hellwag,[11] and Storr.[12]

Dodart's scientific achievement consists primarily in having shown that the trachea—like the wind-chest of an organ—has only the function of conducting the air from the lungs to the larynx, whereas since Galen it had been believed that it served a particular function in voice-production. Dodart also pointed out the importance of the widening and narrowing of the vocal folds, but overestimated it. It was only forty years later that Ferrein could show that the tension of the vocal folds is of decisive importance for voice—although he in turn neglected the widening and narrowing of the vocal folds. These physiological investigations were carried out overwhelmingly in the field of practical medicine—by reason of the division of the faculties. However, medicine too at the beginning of the 18th century—and German medicine especially—was simultaneously under the influence of Leibniz and Christian Wolff. It was still overwhelmingly orientated towards completing its liberation from speculative metaphysical connections. Names like Newton, Bernoulli, Euler, Franklin, Volta, and Galvani indicate the main achievements in mathematics and physics. These sciences were not only enlisted in the service of medical physiological research as auxiliary sciences, but were also a final model for it. Moreover, at that time, most of the outstanding doctors were preoccupied with general scientific questions, especially with chemistry and physics—for the traditions of the schools of the iatrophysicists and iatrochemists in 18th century medicine were by no means lost.[13]

Their relations to Dodart, Ferrein, and the German phoneticians deserves examination. For here it is a matter of formulating physiological problems—and that is particularly true of Haller's classic investigations into breathing:[14] it was not the defining and distinguishing of languages that was of interest, but the production of voice—and that indeed (as is proper for physiological investigations) in connection with the question of the production of voice in animals.

In 1781 C. F. Hellwag published his vowel-triangle.[15] All such vowel-charts are based, as Jespersen emphasizes, on a "general similarity of sounds."[16]

So far as we can determine, the first to express the idea of a vowel-polygon in Germany was the Göttingen physicist G. C. Lichtenberg,[17] who in the 1760's or 1770's attempted to represent the vowels "like Mayer's color-triangle."[18] Even earlier, painters and other pigment experts had attempted—quite independently of the order of spectral color—to represent synoptically the whole known realm of color. The earliest, still quite primitive attempt is seen in the color-chart published by J. Brenner in Stockholm in 1680. There is a small advance—nine years later—in the work of R. Waller.

> Waller separated . . . the blue pigments from the yellow and red ones, and thus only retained mixtures of such pairs. He therefore overlooked the fact that a complete table would exist only when he had placed all pigments on every side. He had also foreseen only one mixture of each pair (in equal parts), since otherwise the table would have been too extensive. Then he would in any case not have managed with the two measurements of the table. . . . The difficulties, which were here evaded but not overcome, were solved two generations later by the distinguished mathematician Tobias Mayer in Göttingen (1745). He began with the three-color-theory, and produced from the three basic colors yellow, red, and blue, in gradations of 1/12 each, first all twofold then all threefold mixtures, so that all possible combinations within the twelve grades appeared. These he arranged in a triangle, in whose corners are the three pure colors; the sides are formed by the twofold mixtures and the interior by the threefold ones. Besides this triangle, he constructed a number of others, which were blended in the same way from basic colors, which had received a measured addition of white or black. He believed that in this way he had provided for all imaginable colors. This was not the case, since for his basic colors he had no ideally pure-colored pigments at his disposal. Moreover, in the middle of the first triangle appeared muddy blends which reappeared in the other triangles. Mayer did not publish his work, perhaps because he had noticed this inadequacy himself; it was published after his death by Lichtenberg.[19]

The indirect connections of the method of comparative linguistics with that of comparative anatomy had a much deeper effect than those direct connections of phonetics with voice physiology on the linguistic research of the previous century, especially on the endeavors of the Neogrammarians and at the same time on the concept formation of the beginning experimental phonetics.

To throw light on the connections and especially on the unique role of intermediary that Goethe played at the end of the 18th century, it is necessary to treat the most significant stages of biological research, from Haller's physiology via Linneaus' systematic achievements to the beginnings of evolutionary history and comparative anatomy, in more detail than is usually found in historical expositions of phonetics and linguistics.

In the work of Albrecht von Haller the enormous knowledge accumulated since Vesalius in anatomy and physiology was collected into one large structure. This was really the first time since Galen that such a compilation had been made, and the circumstances under which Haller was active were in many respects similar to those which Galen faced in the 2nd century A.D.

In antiquity research in anatomy and physiology received an enormous impetus with the great Alexandrians and, over several centuries, led to the assembly of a really astounding stock of facts in these fields. Through Vesalius (1514–1564), these two subjects, the foundations of medical theory, were roused to new life after almost 1,500 years' dormancy and discovery was heaped on discovery. As before Galen, Graeco-Roman medicine had, in the flickering conflict of the different schools (Dogmatic, Empiric, Methodic, etc.), attacked the basic problems of medicine from the most varied angles and had attempted—although briefly in a biassed way—to solve them, not without some progress; so, from Paracelsus on, similar battles were fought between the Iatrochemists, Iatrophysicists, Hippocratics, etc. and in both epochs finally led on the one hand to a reaction, in the sense of a preference for the purely practical-clinical viewpoint. On the other hand, however, as in Galen's time so also again before Haller, the great but unintegrated advances in the cognition of nature pressed towards synthesis by a single outstanding mind according to a unified basic thought. And as in the 2nd century A.D. Galen had undertaken this task, so in the 18th century did Haller . . . Haller's most important writings concern anatomy and especially physiology—here also there is an affinity with Galen. Physiology lacked any kind of unified, synoptical treatment before him, despite the numerous outstanding individual achievements (Harvey and his successors, for instance). It was a collection of many facts. An outward expression of this was that since Galen's *On the Uses of the Parts of the Body of Man*, not a single real text-book of physiology had been written. So much more prodigious, then, appear the two books by Haller, the elementary work *Primae lineae physiologiae* of 1747, and his great handbook of 1757 *Elementa physiologiae corporis humani*. On a par with these are the *Icones anatomicae* and the commentaries on Boerhaave's *Institutiones*. His chief merit lies in replacing the vague concepts, by which one had previously attempted to explain the organic process of movement as the basis of a large number of functions of life, by a scientific fact, and thereby being the first to solve this basic biological problem exclusively on the basis of physical experience, without any recourse to speculation . . . always endeavoring to give his theories a firm scientific foundation by the closest connection with special anatomy and the undertaking of countless experiments. Although from our present-day standpoint much that is incorrect and still more that is imperfect is retained in his physiology, credit is still due to him for having achieved advancement in this science and for having shown the ways in which it could be further elaborated.—While Haller himself was perfectly aware that he had only found a few basic forms of organic life, many of his pupils and followers took his theories as something final, beyond which one could not advance; others again, still firmly entrenched in the views of the 18th century, opposed him, and only comparatively few fully comprehended the spirit of his teachings and tried to advance further in this spirit.[20]

Among these latter are two who have won merit in the investigation of anatomical structure and the function of the larynx: Giovanni Battista Morgagni (1682–1771)[21] and Heinrich August Wrisberg (1739–1808).[22]

Without any connection with these medically-orientated physiological investigations, the first edition of Linnaeus' *Systema Naturae* appeared in 1735, and up to 1766–68 it passed through twelve editions.[23] Wotton's work *De differentiis animalium* of 1552 reopened the period of systematic zoology, after a 1,500 year interruption, which then found its most brilliant representatives in Ray and Linnaeus. The work of Linnaeus became the foundation for systematic zoology, since he introduced a sharper division into the system, a definite scientific terminology and clear, concise diagnoses. In the division of the system Linnaeus employed four categories; he divided the whole animal kingdom into classes, the classes into orders, these into genera, the genera finally into species. The term Family was not employed in the *Systema Naturae*.

[However], . . . the way in which he divided the animal kingdom into fundamental groups is, compared with the Aristotelian system, rather to be called a step backward than a step forward. Linnaeus divided the animal kingdom into six classes: Mammalia, Aves, Amphibia, Pisces, Insecta, Vermes. The first four classes correspond to Aristotle's four groups of animals having blood. In dividing invertebrates into Vermes and Insecta, Linnaeus is undoubtedly inferior to Aristotle, who attempted, in part successfully, to establish a larger number of groups. But in his successors, even more than in Linnaeus himself, we see the damage wrought by the systematic method. The diagnoses of Linnaeus for the most part models, which, *mutatis mutandis*, could be applied with little effort to new species. There was needed only some exchanging of adjectives expressing the differences. With the hundreds of thousands of different species of animals, especially of insects, there was no lack of material, and so the stage was set for that mindless zoology of species-making, which in the

first half of the 19th century impaired the reputation of zoology in educated circles. There was a danger that zoology would have grown into
a Babylonian Tower of species-describing, if a counterpoise had not been
created in the strengthening of the physiologico-anatomical side.[24]

The foundation of the Linnean systematization is formed by a definition of Linnaeus', which has played an important role in the history
of the concept of species and thus in the history of comparative anatomy: "Tot sunt species," it is stated in Linnaeus, "quot diversas formas
ab initio produxit Infinitum Ens" [There are as many species as the
Infinite Being produced diverse forms in the beginning]. With this he
built up a conception of species upon the tradition of the Mosaic creation story. This was nothing unusual at that time, when the religious
character of the view of nature still played a decisive role in writings
on natural science for numerous clerics (Linnaeus himself was de-
scended from the Swedish clerical family Ingemarsson; further Eichhorn in Danzig, Goeze in Quedlinburg, Schäffer in Regensburg) have
won themselves an honorable place in the ranks of zoological writers—
a sign that a "reconciliation" had been reached between ecclesiastical
tradition and the study of nature, and that interest in the study of
nature had penetrated wide circles.

But the tradition of the church could not remain permanently a
foundation for research. The very question—originally based on it—
as to the logical value of the systematic concepts species, genus, and
family finally became a principle dominating the whole of zoological
research. The superficial view of the pure empiricist could easily un-
derestimate this question—but it contains the guiding star of empiricism; namely, the task of specifically biological research itself; and so
it finally became the source of the theory of origin of species and of
comparative biological research.

But yet another biological discipline originated during these decades—again relatively independently of the above-mentioned—and
dominated the biological thought of the 18th and 19th centuries: developmental history or embryology. "What Aristotle really knew and
what Fabricius from Aquapendente and Malpighi wrote about the embryology of the chicken did not rise above the range of aphorisms, and
was not of sufficient value to constitute a science. The difficulties of
observation due to the delicacy and minuteness of the developmental
stages were lessened by the invention of the microscope and micro-
scopic technique. Furthermore, the prevailing theoretical views proved
a hindrance. There was no belief in embryology in the present-day

sense of the word; every organism was thought to be laid out complete in all its parts right from the beginning, and only needed growth to unfold its organs (evolutio). This original meaning of 'evolution' is different from that prevailing at present." [25]

Caspar Friedrich Wolff opposed this doctrine with his "Theoria generationis" (1759): he sought to prove by observation that the hen's egg at the beginning is without any organization, and that gradually the various organs appear in it. In the embryo there is a new formation of all parts, an epigenesis. This first assault upon the evolutionist school was entirely without result, chiefly because Albrecht von Haller, the most celebrated physiologist of the 18th century, declared himself against the theory of epigenesis. Only after Wolff's death did his writings find proper recognition through Oken and Meckel.

This is the biological background against which two biological analogies[26] must be seen, which came to be a decisive influence on the linguistics of the 19th century. The comparison of mental events with organic growth and the speculations on the origin of languages moved almost parallel to the biological discussions on the origin of species, although in separate tracks. Both stand in close relationship.

We saw in Helmont that the "Natural," by which he meant the original, was the language spoken by God with the first men in Paradise. Even far into the 18th century the dominant view was that language is innate in man. For without language—so it was argued[27]—man could never have attained development of his reason. He must therefore have received it from the Creator even before the development of his reason. In 1787, according to Kempelen, Court de Gébelin[28] tries "to persuade us by his overflowing eloquence, that not only the organization for language but also language itself is a direct gift of the Creator; that Man, with all his ability, would never have invented a language of his own accord; that all languages are sprung from one original language—and here appears an image which had the deepest influence on the first decades of comparative linguistics—like shoots from one stem; that even those words which man invented are not drawn from his caprice but from the nature of things, and consequently must sound thus and not otherwise." [29]

The anatomical and physiological descriptions of Court de Gébelin must also be understood in the light of these fundamental principles, in spite of the mechanistic sounding chapter headings (e.g., "Analyse de l'instrument vocal, et de son mécanisme pour produire la voix").[30]

But this comparison of the relations of the different languages to an original language, with the shoots which have sprung from one stem, also has its long history. Essentially it is closely related to the comparison of mental events with plant life, which appears in Bodmer's "Kritische Briefe" of 1749 and which Hildebrand and Burdach[31] have referred to: "I had always regarded this nature as a plant, which must indeed be diligently tended if it is to produce tasty fruit, but which puts out its twigs of its own accord."

This comparison is preceded by an image from the sphere of biological processes: it had long been customary to represent relations of descent between organisms, especially between men, in the form of this image of the trunk with its branches—a comparison which naturally is in itself only valid in a restricted sense. Biologically, the members of a family are certainly individuals and are each determined by two germ cells; they are not like the leaves on a tree, of one stock and type.

This image too, like so many, comes from the Bible. Thus it says in Luther's translation of Sirach 40. 15, of 1533, which was then taken over the following year into "Die gantze Heilige Schrifft Deudsch": "Die nachkommen der Gottlosen / werden keine Zweige kriegen / Und der ungerechten wurtzel stehet auf einem blossen felsen" [The children of the ungodly will not put forth many branches / They are unhealthy roots upon sheer rock]. The expression "genealogical tree" was not yet familiar to Luther. Still in 1541 he translates the "biblos geneseos" of Matthew literally: "Buch von de Geburt" [Book of the Birth], and only more recent revisers of the Bible speak of a genealogical tree of Christ.

Devrient has shown how this expression originated: the Romans invented schedules for the representation of relationships of affinity, for legal ends of the laws of inheritance and marriage, to allow an easy survey of the different degrees of affinity. In these, all relatives of the same degree were placed beside or beneath each other.

Besides geometrical figures (circles and parallelograms), architecturally decorated and grouped compilations were employed. When under Hadrian the inheritance law of maternal relatives was introduced, the schedule acquired a symmetrical arrangement of the principal relatives, beneath whom the descendants of the testator were placed in column form. This form then gave the impression of a tree, and in place of architectural ornament, twig and leaf-ornamentation appears. This development, however, took a fairly long time; for in Justinian's *Corpus Iuris* there is still no mention of a "tree." Only in the 7th century do we find it mentioned in Isidore of Seville; he speaks also of "branches," "stump," and "root." The oldest extant drawings come from the 14th

Century. At that time Roman Law was spread by a number of text-books, which regularly contain the "Verwandtschaftsbaum" [tree of affinity]. It also passed into the expositions of German laws, where it is designated "Sippenbaum" [tree of kindred]. In Germany too, this image was employed for centuries, without the contradiction being noticed, that the trunk of this tree (the person under investigation being placed in the middle) shows children, grandchildren, and great-grandchildren down to the root, and parents, grandparents, and great-grandparents up to the foliage! [32]

Besides these legal trees of affinity, the tree appears as an artistic motif in larger genealogical representations in murals from the 12th century onwards. Splendid works of the 15th and 16th centuries then made the application of the tree metaphor to genealogies familiar to all the educated. From the combination of the concepts "Stammtafel" [genealogical table] and "Verwandtschaftsbaum" [tree of affinity] there originated only in modern times the "Stammbaum" [genealogical tree].[33]

These notes were intended only to supplement the investigations of Cassirer on the problem of the "origin of language" [34] and on the development of the ideas of Vico, Hamann (whom Goethe compared with Vico), Herder and Schelling. Cassirer points out the importance of Herder's prize essay of 1772, "Über den Ursprung der Sprache" [On the origin of language], which had an influence not only on Goethe, the Romantic Movement, and the beginnings of comparative linguistics, but also directly on the founder of experimental phonetics, Wolfgang von Kempelen.[35] The influence of Herder on Goethe is of such decisive importance for linguistics and phonetics because Goethe's natural philosophy and nature-studies played a powerful intermediary role for the beginnings of comparative linguistics. Konrad Burdach has outlined these connections,[36] but even here no account is taken of the essential interest which Goethe took in biological research iself, and which was very much less than is commonly supposed, at least in Germany.

To show this, however, it is necessary to describe, at least in a few words, the beginnings of comparative anatomy.[37] Through the industry of two generations a rich store of anatomical facts had been gathered since the appearance of Linnaeus' *Systema Naturae*, and was awaiting elaboration, which was begun by the great comparative anatomists at the end of the 18th and beginning of the 19th centuries.

When the various animals were compared with reference to their structure, there was obtained at a series of important fundamental laws,

particularly the law of the correlation of parts—"balancement des or-
ganes"—and the law of the homology of organs. The former established
the fact that a relationship of dependence exists between the organs of
one and the same animal, that local changes in one single organ also lead
to corresponding changes at some distant part of the body, and that
therefore from the nature of certain parts an inference can be drawn as
to the nature of another part of the body. Cuvier particularly applied
this principle to reconstruct the form of extinct animals from their pa-
laeontological remains.—The theory of the homology of organs became
still more important. In the organs of animals, a distinction was drawn
between an anatomical and a physiological character; the anatomical
character is the sum of all the anatomical features as they appear in form,
structure, relative position, and mode of connection of organs; the physiological character is their function. Anatomically similar organs in
closely related animals will usually have the same functions; for instance,
the liver of all vertebrates has the function of secreting gall: here anatomical and physiological characteristics go hand in hand. However, this
is not necessarily the case; rather it may happen that one and the same
function can be performed by anatomically different organs; for example, the respiration of vertebrates is carried on in fishes by gills, in mammals by lungs. Conversely, anatomically equivalent organs may have
different functions, as the lungs of mammals and the swim-bladder of
fishes. Similar organs can thus undergo a change of function from one
group of animals to another; the hydrostatic apparatus of fishes has become the seat of respiration in the mammals. Organs of similar function—
physiologically equivalent organs—are called "analogous"; on the other
hand, organs of similar anatomical nature—anatomically equivalent organs—are called "homologous." It is the task of comparative anatomy to
discover the homologous organs in the various divisions of animals and to
follow their changes in form and structure which are conditioned by a
change of function.[38]

The most outstanding representative of comparative anatomy was
Georges Dagobert Cuvier[39] (1769–1832), a graduate of the famous
Karlsschule at Stuttgart. His investigations, apart from his epoch-
making researches upon the structure of the molluscs, extended to the
Coelenterates, Arthropods, and Vertebrates, living as well as fossils.[40]
He gathered his extensive practical knowledge of the structure of ani-
mals into his four large works: *Recherches sur les ossements fossiles*,[41]
Le Règne animal distribué d'après son organisation, [42] *Leçons d'anat-
omie comparée*,[43] and *Histoire des sciences naturelles*.[44]

Of really epoch-making significance was his little pamphlet "Sur un
rapprochement à établir entre les differentes classes des animaux," in
which he founded his famous theory of types and which in 1812 introduced a complete reform of systematization. This classification of Cu-

vier's which has become the starting point for all later divisions, is distinguished by the fact that it combines the classes of mammals, birds, reptiles, and fishes into a higher grade under the name of "Vertebrata," introduced by Lamarck; the so-called "invertebrate animals" were divided into three similar grades of equal standing with the vertebrate animals: Mollusca, Articulata and Radiata. Cuvier called these units above the classes provinces or chief branches *(embranchements)*, for which Blainville later introduced the name "types."—Still more important, however, are the differences which appear in the structural basis of the system. Instead of employing a few often external characteristics in the classification, as previous taxonomists had done, Cuvier based his descriptions on the totality of internal organization, as expressed in the relative positions of the most important organs, especially the position of the nervous system. "The type is the relative position of parts" (von Baer). Thus for the first time comparative anatomy was employed in the construction of a natural systematization of animals. Finally, the theory of types established a completely new conception of the arrangement of animals. The prevailing view that Cuvier encountered was the theory that all animals form a single series, from the lowest infusorian to man; within this series, the position of an animal was determined solely by its degree of organization. Cuvier, on the other hand, taught that the animal kingdom consisted of several coordinated units—the types, which exist independently side by side and within which there are higher and lower forms. The position of an animal is determined by two factors: first, by its membership of a type, by the "structural plan" which it represents; second, by its degree of organization, by the stage to which it belongs within its type.[45]

Parallel with Cuvier's investigations appeared the first works of palaeontology,[46] through which the copious store of extinct animals deposited as fossils was made accessible to research.

By strange fantasies the fossils, being inconvenient for study, were for a long time kept outside the pale of scientific research: They might be sports of nature, it was said, or remains of the Flood, or influences of the stars on the Earth, or products of an *aura seminalis*—a fertilizing breeze that, on overtaking organic bodies, led to the formation of animals and plants but that, when straying onto inorganic material, produced petrifactions. Such confused speculations—already attacked by Leonardo da Vinci, Hooke, Buffon, and other unbiassed men—were finally ended with Cuvier's establishment of scientific palaeontology. Cuvier proved beyond a doubt that these fossils were the remains of prehistoric animals. Just as the formation of the earth's crust from successive superimposed layers made it possible to distinguish different periods in the earth's history, so palaeontology taught us how to recognize the different periods in the plant and animal life of our globe. Each geological age was characterized by particular fauna peculiar to it; this fauna differed the more from that now living, the older the geological period to which it belonged. These generalizations led Cuvier to his cataclysm theory; that a great revolution brought each geological period to an end, destroying all life. As to how the earth was populated again after the termination of a cataclysm Cuvier expressed himself very cautiously. Since he advocated the theory that the species were constant and created by a Highest Being, it was a short step to the explanation that after each revolution a new fauna was created by God. This explanation, definitely expressed later by the palaeontologist d'Orbigny, probably corresponded to Cuvier's views; however, the latter also reckoned with a further possibility, that some parts of the earth's surface were spared by the catastrophes and that from them a residue of organisms, which were preserved, spread anew over the earth. Since on more careful reflection neither conception offers a satisfactory scientific explanation, it is conceivable that already in Cuvier's time outstanding zoologists had given up the theory of species constancy and declared themselves for the theory of descent or evolution.[47]

However, it is an error which has maintained itself in Germany since Goethe's lifetime that he was one of these men. One must realize particularly that Goethe's nature studies—despite the large place which they occupy in his life and writings (the scientific works fill 14 volumes of the Weimar edition, not counting the letters and diaries which deal with this subject)[48]—are not research in the sense in which it was conducted even in his time. It is simply not historical to prefer Jena professors, who were dependent on him,[49] as guarantors for his importance as a scientist to the general verdict which zoologists and anatomists[50] who were independent of him already gave at that time. This is true of Goethe's scientific works on the metamorphosis of plants and on the famous intermaxillary bone,[51] as also of his taking sides in the well known Academy conflict between Cuvier and Geoffroy Saint-Hilaire.[52] His natural philosophy would have prevented him from appraising Cuvier's merits, even if the latter's works had been better known to him than they were.[53] And what drew him to Geoffroy—at least after the latter's turning to natural philosophy from 1819 onwards—was a view of nature markedly similar to his own, but which had already at that time long outgrown comparative anatomy. The ultimatum, which Goethe as minister of state presented to Lorenz Oken[54] and which finally caused the latter to resign his professorship, must also at least be taken into account here.

In judging Goethe's attitude to the natural sciences, a whole series of factors must be considered.[55] First and foremost is his well-known antipathy to Newton. Certainly it required courage then to make an assault on this hero of exact research. But with what arguments he did it! All his statements in this conflict show that he was entirely unable even to appreciate Newton's achievement, much less to refute it. It was popular for a time to describe certain changes in modern physics as having been anticipated by Goethe, and to construct connections with his arguments against Newton. It must be said here that in any case, with this type of "historical" construction, everything can be brought into relation to everything else; and further, that modern physics has developed from classical physics and retains not only this historical connection but also its essential connection with it.

It did not develop from Goethe's color-theory, however.[56] The value of "presentiments" should not be disdained. However, science may demand proofs. "Making plans is often a vain business" (Kant). And one should not forget that the whole technology of the 19th and 20th centuries has developed from classical—Newtonian—physics, and that only the most recent questions are made under presuppositions for which those of classical mechanics are not adequate. Goethe's natural philosophy prevented him from understanding it, as did his deeply-rooted aversion to instruments and numbers. Not only Newton's color theory but also the whole of physics since Galileo is based on them.[57]

Goethe's attitude to mathematics is well known. Here only a less-known letter of F. H. Jacobi to Köppen, of 24th July, 1805, will be cited: "I believe he [Goethe] would gladly have demonstrated to me that he could adopt all my facts into his system, but that mine lacked certain facts of his system. Once he became almost angry, when I made it too clear to him that, as Pascal says, *ce qui passe la géométrie, nous surpasse* [What passes geometry surpasses us], and that consequently a speculative theory of nature after the modern fashion could only be a chimera." [58] This correspondence, which appeared during his lifetime, was discussed by Goethe accordingly: "Jacobi neither knew nor cared at all about nature, indeed he stated clearly: it hides his God from him. Now he believes he has triumphantly proved to me there is no natural philosophy; as if the external world did not reveal everywhere day and night the most secret laws to *him who has eyes* [our italics]. In this consequence of the infinitely manifold I *see* [Goethe's italics] God's handwriting in the clearest possible way." [59] And as confirmation he adds for his astonished readers a "translation" of the famous passage from Dante's *Inferno* (XI, 98), which he ends with the words: "Natural philosophy [in the Weimar sense, of course—author's comment] is the grandchild of God." On the other hand, he writes, also in the criticism of this correspondence, "the Swiss shakes his head over the Saxon, the Viennese over the Berliner; concerning the real matter in question, one knows as little as the other; with few exceptions they all dance at the wedding but none of them has seen the bride." Meanwhile, outside Weimar, the principles of research were developing; namely, the principles of comparative biology, of geology and palaeontology, of geography, of comparative linguistics and philology, and of history.

It can further be shown that he had read, long before and often,[60] the general ideas of the relationship of all living creatures, which are credited to him—only less in works of the natural scientists (who were cautious on this point, since at that time the evidence for it was not decisive enough) than in those of philosophers. In 1777 Frau von Stein had written to Knebel: "Herder's book makes it probable that we were once plants and animals. What nature may now conjure out of us further, will probably remain unknown to us. Goethe is now pondering very deeply on these matters." [61]

In the first impression of Kant's *Critique of Judgment*, which appeared in the spring of 1790 and which was in Goethe's library, he marked the critical passages with his own hand, and already on October 6 of the same year, Körner wrote to Schiller that Goethe "has found food for his philosophy in the critique of teleological judgment." In 1817 Goethe was delighted by "the passages in the old copy [of the *Critique of Judgment*] which I marked then." [62] And which were those passages? One of those which is critical in this context is at the beginning of the theory of method of teleological judgment (§80); it runs as follows:

> It is praiseworthy to employ a comparative anatomy and go through the vast creation of organized beings in order to see if there is not discoverable in it some trace of a system, and indeed of a system following a genetic principle. For otherwise we should be obliged to content ourselves with the mere critical principle—which tells us nothing that gives any insight into the production of such beings—and to abandon in despair all claim to *insight into nature* in this field. When we consider the agreement of so many genera of animals in a certain common schema, which apparently underlies not only the structure of their bones, but also the disposition of their remaining parts, and when we find here the wonderful simplicity of the original plan, which has been able to produce such an immense variety of species by the shortening of one member and the lengthening of another, by the involution of this part and the evolution of that, there gleams upon the mind a ray of hope, however faint, that the principle of the mechanism of nature, apart from which there can be no natural science at all, may yet enable us to arrive at some explanation in the case of organic life.[63]

The passage then continues—although no longer marked by Goethe:

This analogy of forms, which in all their differences seem to be produced in accordance with a common type, strengthens the suspicion that they have an actual kinship due to descent from a common parent. This we might trace in the gradual approximation of one animal species to another, from that in which the principle of ends seems best authenticated, namely from man, back to the polyp, and from this back even to mosses and lichens, and finally to the lowest perceivable stage of nature. Here we come to crude matter; and from it and the forces which it exerts in accordance with mechanical laws (laws resembling those by which it acts in the formation of crystals) seems to be developed the whole technic of nature, which—in the case of organized beings—is so incomprehensible to us that we feel obliged to imagine a different principle for its explanation.

Here the *archaeologist* of nature is at liberty to go back to the traces that remain of nature's earliest revolutions, and, appealing to all he knows of or can conjecture about its mechanism, to trace the genesis of that great family of living things (for it must be pictured as a family if there is to be any foundation for the consistently coherent affinity mentioned). He can suppose that the womb of mother earth as it first emerged, like a huge animal, from its chaotic state, gave birth to creatures whose form displayed less finality, and that these again bore others which adapted themselves more perfectly to their native surroundings and their relations to each other; until this womb, becoming rigid and ossified, restricted its birth to definite species incapable of further modification, and the multiplicity of forms was fixed as it stood when the operation of that fruitful formative power had ceased. Yet for all that, he is obliged eventually to attribute to this universal mother an organization suitably constituted with a view to all these forms of life, for unless he does so, the possibility of the final form of the products of the animal and plant kingdoms is quite unthinkable.[64]

On this the following observation by Kant:

An hypothesis of this kind may be called a daring venture on the part of reason; and there are probably few even among the most acute scientists to whose minds it has not sometimes occurred. For it cannot be said to be absurd, like the *generatio aequivoca*, which means the generation of an organized being from crude inorganic matter. It never ceases to be *generatio univoca* in the widest sense of the word, as it only implies the generation of something organic from something else that is also organic, although, within the class of organic beings, differing specifically from it. It would be as if we supposed that certain water animals transformed themselves by degrees into marsh animals, and from these after some generations into land animals. In the judgment of plain reason there is nothing a priori self-contradictory in this. But experience offers no example of it. On the contrary, as far as experience goes, all generation known to us is *generatio homonyma*. It is not merely *univoca* in contradistinction to generation from an unorganized substance, but it brings forth a product which in its very organization is of like kind with that which produced it, and a *generatio heteronyma* is not met with anywhere within the range of our experience.[65]

In the following paragraphs, in which again several passages were marked by Goethe, Kant quotes Johann Friedrich Blumenbach's book *Über den Bildungstrieb*, the first impression of which appeared in 1781, the second in 1789.[66]

Goethe's *Metamorphosis of Plants* (which moreover does not consider the roots as well, since they are under the earth, invisible to the eye!) was written in 1790; in 1795 and 1796 he put on paper the draft of his general introduction to comparative anatomy.

But besides his attitude to physics, and the sources from which he drew his views on biology, one must take into consideration that he took virtually no interest in the really decisive biological researches of his time—in Caspar Friedrich Wolff, in Cuvier—and at the end we must notice that one of the results of the development of biology was that it too, as biophysics, biochemistry, and biological statistics, finally became a discipline that can no longer dispense with experiment and statistics.

It is often forgotten that in 1809 Lamarck's "philosophie zoologique" had already appeared, containing a clearly thought out theory of evolution, among whose founders Goethe is often erroneously counted. This theory made an essential advance, in the same year as Goethe's death, through the final establishment of developmental history—and that through Karl Ernst von Baer's classic work: *Die Entwicklung des Hühnchens. Beobachtung und Reflexion.* In this Baer confirmed Wolff's doctrine of the appearance of layer-like processes from which the organs arose, established the so-called germ-leaf theory, and came to the conclusion that each animal type possesses not only its peculiar structural plan, in the sense described by Cuvier twenty years previously, but also its peculiar course of development. He thereby became the creator of comparative embryology, which offered powerful support to the idea of comparative anatomy.

During the last decades of the 19th century, physiological investigation of the organization of animal forms took its place beside morphological investigation, and won more and more ground from the latter. Its most important—and quite un-Goethean—means of research are the experiment and methodical breeding, which at the same time provide the most important auxiliary means of investigating in greater depth, by statistical methods, the variability of organisms (e.g., the work of Quetelet, Pearson, Mendel, Galton, and Johannsen).

However, the question as to what significance Goethe's biological pursuits had for the progress of zoological or botanical research is not properly formulated. Taken all in all, soberly weighed and seen historically in the sense in which he himself said of the *Critique* in 1815—that it could be nothing more "than the observation of the various effects of time"—his contribution could only turn out to be negative. He did not further pure research; but rather—by his speculative natural philosophy and the influence which he as a person and as a minister of state possessed—he retarded it.

It must rather be asked, if one understands "the effects of time" positively: what significance have Goethe's nature studies for his own life and life-work? Only then is it apparent that they return magnificently to language and the intuition linked with it, just as they set out from language and intuition.

However, his natural philosophy had, in the context of his whole activity, an unexpected, fruitful effect on the beginnings of comparative linguistics, on which at the same time Kant and Humboldt also had a formative influence.

The theory of language played only a very subordinate role for Kant—and consequently in Neo-Kantianism also—at least until the very latest period. So much the more important is the intermediary role which he played. The *Critique of Aesthetic Judgment* influenced Goethe, but, since their minds were more alike, even more decisively Schiller.[67] Schiller, who had also studied carefully the *Critique of Pure Reason*, passed Kantian philosophy on to Wilhelm von Humboldt. In 1791 Schiller wrote to Körner: "I am carried away by the brilliant ingenious content of his *Critique of Judgment*, of which I have acquired a copy. It has aroused in me the greatest desire to familiarize myself little by little with his philosophy"; in January, 1792; "I am now studying Kantian philosophy with the greatest zeal"; and in October of the same year, again to Körner: "I am now up to my ears in Kant's *Judgment*. I shall not rest until I have mastered this material and it has become something under my hands." [68] Of course it was principally the Kantian theory of beauty that absorbed him and with which he was coming to terms—but quite in the sense of the *Judgment*: always with an eye to the theory of the organism. In February, 1793 he wrote whole treatises on Kantian philosophy to Körner.

Wilhelm von Humboldt had already known Kant since his time in

Frankfurt and was also preoccupied with Kant and Plato in 1791; but Kant's full significance only became clear to him—after he had passed through the school of Friedrich August Wolf[69] and absorbed Herder—when he came into contact with Schiller and Körner in 1794.[70]

At this point the question must again be raised: what is the nucleus of the Kantian doctrine, the nucleus especially of the Copernican turn, the essence of Kant's transcendental method, if one wishes to form a picture of the specific achievement of Wilhelm von Humboldt? This "turn" at the beginning of Kant's critical period has a systematically almost direct connection with Copernicus himself, to whose spirit Bruno bore witness on the threshold of the 17th century by his death. It is the spirit of Copernicus, for which Galileo also lived and suffered and in which he investigated. "To proscribe Copernicus," he declared, "is to proscribe science itself. The problem of the true organization of the world is the most important in itself and also for the light which it throws on all the other natural sciences. And there is such an organization, and it exists in a unique, true, and so necessary way that it cannot be other than it is." [71] When in 1609 Galileo trained on the sky the telescope invented or copied by him, discovered the "Medicean stars" (as he called the satellites of Jupiter), and pointed out a central body, with the satellites circling it—then Copernicus' view, together with Kepler's commentaries, received its most convincing empirical support; then he already possessed a knowledge of the laws of free fall—i.e., he had then created the two pillars for Newton's theory of gravity. It has been explained before what the discovery of the laws of free fall meant for cognition theory [72]: in them, a new type of experience was created by a new combination of mathematical construction and sense perception: the experience of the exact sciences based on experiment; that is, a cognition of nature in comparison with which all other knowledge of nature remained subjective.

It was with this that Kant started: there is only *one* exact cognition of nature (whose temporal beginning may lie with the senses): that is exact natural science. No room remained beside it for a philosophy of nature, for this could only be unguaranteed speculation which must yield step by step to natural science (i.e., to the physical experiment)— if this philosophy does not restrict itself, like metaphysics, to the discussion of a transcendental object, beyond all experience. But a new task develops out of this state of philosophy: the investigation of the questions: on what then is the particular achievement of exact scientific research based, how does it differ from the subjective knowledge of natural circumstances on the one hand, from pure mathematics on the other; what presuppositions will have to be made, so that the experiment confirms the theory for which it is undertaken? In other words: from a philosophy of nature, Kant developed a theory of cognition of nature; that is, a theory of physics—that is, of physical research. The great physicist Helmholtz, in his famous Königsberg speech of 1855, said, "Kant's philosophy did not aim at increasing our knowledge quantitatively by pure thinking, for its highest thesis was that all cognition of reality must be drawn from experience; it only aimed at examining the sources of our knowledge and the degree of their justification—a task which will always remain for philosophy and which no age can evade with impunity." [73]

What Kant with his new way of seeing these circumstances founded—something completely new—was the indissoluble connection (indissoluble especially for every cognition) between the physical judgment on the one hand and the object of nature on the other. We only mention here in passing that he discovered a completely new theory of judgment, which was only perfected in the modern psychology of thought. He showed that only two ultimately confirmed attitudes towards nature are possible: that of the physicist, who makes judgments about nature on the basis of experiments, and that of the epistemologist who examines the type of this judgment and therefore asks, for example, what thoughts of the spatial, temporal, or causal context of the phenomena enter the thought of the experiment, as these ideas are not a consequence of the physical judgment nor can they be proved by it, like the law of nature itself; these thoughts logically (however not temporally) precede the natural law. Hereby the physical experiment now becomes, vice versa, the proof of these thoughts, since one can say: if one can experiment only under these or those hypotheses and in fact does experiment successfully, then these hypotheses are obviously suitable for the cognition of nature; that is, then they are proved both as conditions for the judgment *about* nature and also as hypotheses of nature itself: these ideas prove or, if one prefers, create the specific connection between judgment and nature. And this is what Kant called a transcendental proof.

However, even he in the "Judgment" did not remain faithful to this theory which is here deliberately presented in rather Neo-Kantian terms. The external cause lay close at hand: for Kant there existed only two real and exact sciences; namely, mathematics and physics; and in the theory of experience he had put both sciences into a new cognitive

relation deriving from Galileo, and he distinguished both in principle
(that is, in their direction). Biology began only now to develop a task
of its own; a comparative anatomy and physiology or developmental
history (this development took place particularly in France). The
"Judgment" shows what a lively interest Kant took in these endeavors.
But he was far from ranking these investigations with physics and
mathematics, as two sciences with their own tasks and their own forms
of judgment, their own concepts of proof and their own hypotheses.
He would have had to risk damaging the foundations of his own structure had he done this. And the same is true even more of the work of
art—which he primarily understood to be the visual work of art. He
certainly knew Winckelmann; but a history of art, in the sense in which
the 19th century developed it, did not yet exist even as an idea. There
was no history of literature in the modern sense; historiography, which
was to be created a little later by Niebuhr and Savigny, by the Grimms
and Gervinus, did not yet exist. What could he do but develop a theory
of the organism, a theory of art, which could not yet be oriented on a
judgment about organisms and a judgment about works of art (that is,
on biology and history)? And so he looked for something that created
intellectually, that was hidden in the creation of organism and work
of art, that both bears the whole in its parts and expresses an infinite
correlation of all parts with one another.

It was this, albeit poorly expressed, which held Schiller's interest in
the *Aesthetic Judgment*[74] and it was this also that Wilhelm von Humboldt saw in it.[75] Through Friedrich August Wolf, Humboldt obtained
a new attitude to antiquity, especially to Plato, and hence also to the
Platonic dialogues on language. In 1795 Wolf's classic "Prolegomena
ad Homerum" appeared, which prepared for the great Homer edition.
And Wilhelm von Humboldt had passed through Herder's school: he
knew the latter's academic dissertation on the origin of language; he
knew that the personality is to be understood as the essence of a particular national spirit. Then he learned to take Kant and the *Judgment*
seriously; what was more obvious than to create a philosophy of language, just as Kant had created a philosophy of art and of the organism—and what was more obvious than to weave aesthetic and biological
tendencies into this new theory? And that is Humboldt's philosophical
feat, or rather the direction in which it was performed systematically
and historically and in which it acted on the whole century following,
up to the present day.

In September and in November of 1795, Wilhelm von Humboldt

wrote to Schiller that language must be comprehended as an organic whole and in the "context of the individuality of those who speak it"; one must "find the categories in which one can place the idiosyncrasies of a language, in order to find the way to describe a particular character of any language." It is this that Humboldt had learned to see, through Wolf, Herder, and Kant.

Not in the consequence of the transcendental method of the *Critique of Pure Reason* but in analogies to the *Critique of Judgment* did he now place the stress on the particular spontaneity of language, which consists in grasping what is said. And thus the philosophically important property of language seemed to him to be that it is not a created, handed-down tool; but it operates and is to be operated, and in this dimension lie the roots of both its object-reference and its subject reference and of its own variability. At the same time, in this analogy to the *Judgment*, lies the systematic root for all later analogies of linguistics to comparative anatomy.

Schiller's letters introduce the history of this theory, which was influential both in linguistics and philosophy. Here one must keep in mind that the beginnings—which are decisive because they show the direction of further development already laid down—lie before the beginnings of comparative linguistics, before the development of the historical method and of philology as its most important auxiliary science. Indeed, they lie even before the first relatively great works in which comparative anatomy found its own task and method, in which Wilhelm von Humboldt had already been interested by his brother. In the same year that Wilhelm joined Schiller, Alexander was led by his scientific interests to Goethe.

To the last decades of the 18th century belong the linguistic observations (begun anew two hundred years after Filippo Sassetti) on the relationship of Sanskrit to the well-known European languages.

In 1767, in a learned dissertation on the relations of Sanskrit to Latin, which he submitted to the Institut de France, the French Jesuit missionary Coerdoux pointed out the similarity of a striking number of Sanskrit words to Latin words. Besides this he undertook grammatical comparisons for the first time: unusual for the 18th century, since the opinion prevailed that languages are distinguished only by their vocabularies, but that all have more or less the same—naturally Latin—grammar. Deviations were then considered nothing more than bad Latin or incorrect grammatical formations, as the case may be: vulgarism, which did not possess scientific significance and did not deserve consideration (just as until recently the opinion was widespread that there is *one* universal alphabet and that the alphabets of individual languages are portions of this one total stock).

William Jones—who had been a judge of the Supreme Court in Calcutta since 1783—characterized it decisively two decades later: "The Sanscrit language, whatever may be its antiquity, is of a wonderful structure; more perfect than the Greek, more copious than the Latin, and more exquisitely refined than either; yet bearing to both of them a stronger affinity, both in the roots of verbs and in the forms of grammar, than could have been produced by accident; so strong that no philologist could examine all the three without believing them to have sprung from some common source, which perhaps no longer exists." [76] Here it is stated for the first time that the proof of a relationship must be obtained by refuting the suspicion of a "chance" similarity or identity of certain linguistic quantities. This is what had happened previously. Sassetti and Coerdoux had referred to the number of similar words exceeding all reasonable expectation (Coerdoux also to grammatical similarities). Jones employed the degree of similarity—both of roots and of grammatical forms—as the criterion. Over and above the number of similar words and forms, the concept of similarity itself is discussed and weighable principles of similarity are sought: not the number of similar words, but their *degree* of similarity, argues against their chance-character. That is, the stress has shifted from the statistical concept of frequency to the linguistic concept of the degree of similarity, without the latter being previously altogether lacking or the former subsequently quite disappearing.

In 1787 the Königsberg philosopher and professor of fiscal sciences Christian Jakob Kraus reviewed and criticized in the *Allgemeine Literatur-Zeitung* the first section of the Empress Catherine II's great universal glossary, just completed by the scientist P. S. Pallas.[77] In this review he says: "One is justified, with good reason, in deducing from individual similarities—for example, of grammatical form, word-order, connection of the verbal material of two languages—a further agreement. . . . For the grammatical method of his language clings to a man more strongly than its material . . . Thus a short comparison of the characteristic features of the structure of the languages can serve very well to point out safer paths to the business of word comparison, which is as laborious and complicated as it is precarious and tempting. . . ." [78]

This insight was elevated to a principle by the Hungarian philologist Samuel Gyármath twelve years later,[79] and independently of him the

Spanish Jesuit Lorenzo Hervás y Panduro drew attention to the importance of grammatical structure in language comparison.[80] However much the stress is placed on linguistic arguments, they are always accompanied by two statistical problems: that of "predominant usage," and the related one of deduction from the examined texts to the language or to texts not yet examined, and this involves statistics. But these questions were not followed up.

6. PHONETICS OF THE NINETEENTH CENTURY

After the turn of the century, Friedrich Schlegel studied Sanskrit in Paris, filled with the hope that the study of old Indian writings would provoke a revolution similar to the revival of the study of Greek at the time of the Renaissance.[1] At the same time he was fascinated by the question of the relationship of the languages which W. Jones had already considered. His teachers Hamilton and Langle referred him to Jones. "The old Indian Sonskrito," Friedrich Schlegel wrote, "that is, the educated or perfect—also Gronthon, that is, the language of writings or books, has the greatest affinity with the Roman and Greek and also with the German and Persian languages. The similarity lies *not only* in a *great number of roots* which it has in common with them, but it extends to the innermost *structure and grammar*. The agreement is therefore not a *chance one*, which could be explained by mixing, but an essential one, which indicates a common descent." [2] With this Friedrich Schlegel rejects *expressis verbis* the conclusive power of the number of observations of roots—because even a large number of such similarities "could be explained by mixing"—and points out the weight of a similarity of "innermost structure and grammar" that proves an "essential similarity"—i.e., one that indicates "common descent."

Since with similar roots one could not decide whether their similarity was explainable by mixing or indicated common descent, even large—unexpectedly large—numbers of similar roots were understood by Schlegel as possibly fortuitous. Here one sees the justification of Emanuel Czuber's cautious formulation, that a large number of sciences, "besides their specific means of research—occasionally!—make use *also* of the common means of statistical method." [3] It is thus not his opinion that statistics could replace the "specific means" of the different sciences and in this way create a kind of unified science. Even these first sentences from Schlegel show rather the precedence of linguistic arguments over statistical ones—without thereby foregoing estimations. After he has put the real "affinity of roots"—as against etymological affectations—on a firmer footing in the following chapter, there follows in the third chapter[4] the now famous sentence: "That decisive point, which will illuminate everything here, is, however, the inner structure of languages or comparative grammar [which Schlegel puts in place of the old "universal grammar"], which will give us completely new insights into the genealogy of languages in a way similar to that by which comparative anatomy has thrown light on the higher natural history." [5]

In Paris Schlegel also became acquainted with this comparative anatomy—at first in the form of comparative osteology. And in the fields of geology, petrography, and palaeontology it was again the question as to the accidental structure of the finds that had to be refuted. Here again the first discussions of the problems go back to the 16th century—back to Leonardo da Vinci: "If the Flood," he writes in 1505, "had carried the shells from the ocean 300 to 400 miles here, they would have been mixed up with other things in a heap [that is, according to the laws of chance distribution]. Instead, we find them . . . all together still. . . ." [6]

And 150 years later Nicolaus Steno[7] speaks of an "inner arrangement of parts," with which he contrasts superficial arrangement: "If a solid body corresponds to another not only superficially, but also according to the *inner arrangement of its parts and particles*, this will also be true with regard to the method and place of generation. From this it follows . . . that those bodies which are dug out of the earth and are *similar in everything* to parts of plants and animals, have originated in the same way and at the same place (that is, in the same element) as the parts of plants and animals themselves." [8]

A generation later, Johann Jacob Baier writes in his *Nürnberger Fossilkunde* [Oryctographia Norica] of 1708: "Moreover, to regard the ammonites not as chance *follies of nature*, but as real shells of shellfish, I see myself obliged for several reasons . . . 1. Not only the whole race, but the individual species too have reliable and *constant characteristics*, so that even if a thousand *individuals* of one species were found, *none is completely lacking these characteristics* [a purely statistical problem, whose relationship with the problem of the "universality of sound laws," discussed particularly in Germany 150 years later, is immediately obvious]. . . . 3. *The work of art of the inner structure*. . . ." [9]

It seems to us that the concept of inner structure derives from this, which here as a work of art, like the clockwork of a watch, is contrasted with chance accumulation.

In 1795 George Cuvier (then 26) became a professor at the central school of the Pantheon in Paris and assistant to Mertrud, the teacher of comparative anatomy at the Jardin des Plantes in Paris. He was the first to begin the scientifically based investigation of fossil vertebrates, and he operated with a statistical concept that can only be understood in its opposition to chance; namely, the *law of correlation* of the organs. To this he had been forced by observations, but he then made this law (as biologically universally valid) the basis of his further investigations: "Fortunately comparative anatomy had a law which, in its practical execution and application, could remove all difficulties: the law of the mutual relationship of forms in living creatures, with the help of which, strictly speaking, each of these creatures could be recognized from each fragment of any of its parts." [10] As so often in the history of the sciences—with Gay-Lussac and Liebig,[11] with Broca and Wernicke, with Cuvier or Geoffroy and Goethe—a Frenchman was the inaugurator and a German gave the productive answer: it was certainly no "accident" that the accountant of the German Federal Diet, Hermann von Meyer, defending the "play" of the "individual phenomena" against Cuvier's law of correlation, discovered the phenomenon of convergence, which restricted the validity of Cuvier's law.

From 1801 onwards, Cuvier's *Leçons d'anatomie comparée* appeared in five volumes. Friedrich Schlegel had become acquainted with them in Paris and he wished the argumentation of comparative grammar to be modelled on this strict demonstration. The scientific treatment of language began in 1816 with Franz Bopp's work *Über das Konjugationssystem der Sanskritsprache in Vergleichung mit jenem der griechischen, lateinischen, persischen und germanischen Sprache*; and in 1818 Rasmus Rask's *Undersøgelse over det gamle nordiske eller islandske Sprogs Oprindelse* followed—the work had been finished since 1815; in 1819 the first volume of Jacob Grimm's German grammar appeared; in 1821, Andreas Schmeller's *Mundarten Bayerns grammatisch dargestellt*, and his *Bayrisches Wörterbuch* from 1827 onwards; in 1835 and 1836, Humboldt's epoch-making works, *Über die Verschiedenheit des menschlichen Sprachbaues* and *Über die Kawisprache*—both published after his death by his brother Alexander; and from 1836 onwards, the *Grammatik der romanischen Sprachen* by Friedrich Diez.

Scarcely any other science has in its beginnings such a preponderance of German research. Berthold Delbrück began his *Einleitung in das Sprachstudium* (in 1893), with Bopp and his views on the origin of inflection. Delbrück shows the dependence of the early Bopp on Schlegel and the latter's Romantic philosophy, particularly its contrast between the inflecting languages that originated "organically," and languages "that instead of inflection have *only* affixes," and whose roots are "not fertile seed" but only like a *"heap of atoms*, which every *wind of chance* can easily disperse or join together"—until the formal renunciation of Schlegel in the *Vergleichende Grammatik*, which appears from 1833 onwards.[12] To the question concerning the general observations from which Bopp makes judgments on events in language, Delbrück gives the now famous answer:

> His general views had a scientific tinge, but under this the old philological priming had not yet disappeared. The love of the scientific manner of expression reveals itself at once, when he tries to describe his method of treating language as against the earlier one. He aimed at a comparative "dissection" of languages: systematic language-comparison is a language-anatomy," it is a question of an "anatomical dissection" or "chemical decomposition" of the body of language, or in another image, of a "physics" or "physiology" of language. The scientific tinge appears very decisively in the very first sentence of the preface to the Comparative Grammar. . . .[13]

If one glances at Bopp's "physical laws," one sees at once that it is here again a matter of statistical problems—in linguistic terms, of course, and therefore naturally with linguistically defined quantities: a matter of the question of the exception to the rule which Grimm designated "Sound Laws" and immediately adjusted to a Romantic circulation theory, "because the spirit of language has completed his course[!]." [14] In view of the experience of the so-called "exceptions," there were only two possible attitudes: to look for the "regular" causes which produce these "exceptions," or to be content that the observed regularities are "usually" or "sometimes" valid—concepts that, again, cannot be comprehended other than statistically.

However, we must examine Humboldt's philosophy of language rather more thoroughly. For it is not only connected with his more strictly linguistic works, but was influential also on Pott and Curtius directly, and had an indirect—but traceable, step by step—influence on phonetics.

First a word on the general consequence of his language theory. The Kantian amalgamation of the organic and aesthetic principles on the one hand, the combination with Wolf's philology and Herder's view of history on the other, led him to a particular, inimitable form of the aesthetic and historical characterization of the individual. This method is inimitable in both senses of the word: it is not only unattainable, it is also unlearnable. It remains an individual act—also in the negative sense of the word: it is not method in the sense of modern research. What was important to him was not the discursive development of a state, the gradual, unending approximation of Galilean science, not the methodical work which extends over generations, but the stylistic definition of a phenomenon, the literary inimitability of the surveying glance. Herein lies the secret of his style, which leads to the deepest depths but on paths which are not rediscoverable. This effect, which even today is felt from every page of a work by Humboldt, was already clearly perceived in his lifetime. Kant wrote of his essay on differences between sexes: "I cannot decipher this treatise, however intelligent the author seems to be." And Körner wrote of Humboldt's essay on Goethe's "Hermann and Dorothea" that one surmised a content in its sentences. And so Humboldt's life's work remains a fragment and programme[15] in the highest style, which—after the development of the linguistic method—resists a Renaissance.

In the sense of the Marburg school of philosophers and with express reference to the Copernican turn of Kantian philosophy and its returning from philosophical treatment of the objects of nature to the epistemological form of their scientific mastery, Ernst Cassirer has placed the particular spontaneity with which we master the world through language as an outstanding symbolic form beside those of science, art, and myth; he characterized the task of philosophy as surveying the structure of these forms.[16]

Now it is inherent—certainly in the sense of the *Aesthetic Judgment*, but not in the sense of the transcendental method of *Pure Reason*—that we must return through philosophy from the object to the act of its comprehension. It was transcendental to make the act of scientific comprehension (the *scientific* method) the pivot of cognition. For in fact it is certainly the spontaneity—to which Kantian philosophy traces the regularity of nature—that is the real structure of order, with which the physicist approaches his objects, and that mathematical physics presupposes when it passes judgment on the regularity of a natural event, basing itself on the methodical experience of experiment.

The transcendental sense of Kantian philosophical research seems to be that cognition theory can make judgments neither on objects nor on non-scientific forms of (biological, aesthetic, or linguistic) activity, but only on *scientific forms of discursive research*. If that is so, it is not the system of "symbolic forms" as with Cassirer but the system of the sciences that is the appropriate subject of cognition theory. Science is therefore, from an epistemological point of view, not *a* form of activity in a system of such forms, not *a* symbolic form beside others, but *the condition* for scientifically defining those other forms of activity, those other "symbolic" forms. That has, since Bopp and Rask, been true of linguistics too, which alone is in a position to determine the linguistic form of human activity, namely of the comprehension of the world; that is, the symbolic form of language. Furthermore, that the comparative-historical procedures followed since Rask and Bopp and the methods of the Neogrammarians are not adequate, that the circle of linguistic procedures should thus be spread wider, has been clearly stated, at least for Europe, by Hans Georg Conon von der Gabelentz and Ferdinand de Saussure before and after the turn of the last century:[17] the procedures of structural linguistics are thus the ones followed today in the study of language.

Certainly, the objective world is mastered through language: nothing can be an object that could not be expressed, at least in principle—even if it be "between the lines"—for that too is characteristic of language.

However, in terms of cognition theory, that is no defined state. It is not "activity" but science that is the starting-point of comparison. And, moreover, this particular "activity of language" can develop only in systems of languages that are handed down. Linguistics alone makes decisions about it and the science of language alone rescues it from vagueness. This activity therefore belongs to the concept of language, which linguistics presupposes—it could otherwise speak neither of structure nor of semantic change nor of morphological change. For this reason scientific philosophy can only come to terms with the "spontaneity of language" by proving it to be one of the conditions of linguistics.

We must have a clear picture of these systematic and historical consequences of Humboldt's linguistic theory, if we wish to judge it. To this belongs his predilection for—mostly unscientific—alternatives that are rooted in the personal decision of the actor but not in the procedure of proof and the advance of research.[18] The alternative most important in our context is the intellectual principle of language on the one hand, and the "natural affinity of the sounds" that are dependent on the organs of speech on the other hand. The form of sounds he calls the sensual side of language—and here one should note that, for him, sound is synonymous to letter. To this belongs further his idea of the mechanical dissection of the corporeal and factual that is contrasted with the understanding of the intellectual part of language.

The aesthetic factor finds expression in his theory of euphony. He is of the opinion that the forms of sounds, as part of the human organism connected with the inner intellectual power, are related to the whole potential of a nation.

His theories are included in full only in his later works on linguistic philosophy, but they already find expression in the "Ankündigung einer Schrift über die vaskische Sprache und Nation," in *Deutsches Museum* of 1812, among whose contributors was Jacob Grimm. In 1823 Grimm wrote to Lachmann: "In the latest volume of the *Berliner Abhandlungen* is a lecture by Wilhelm Humboldt on language-study[19]—do read it, both directions of language and language-study in it seem to me intelligently and excellently developed. Such a thing can console me on the deficiencies in my own works. I am at least on one of the right paths; the spirit which is dormant in the collected material will in time awake or be awakened." [20] Grimm's right path was the procedure which secured the autonomy of linguistics as a science beside physics, mathematics, and comparative biology during the course of the 19th Century.

The analogy initially helping to determine this path was the rapidly developing comparative anatomy, which was already taken as a model by Friedrich Schlegel in 1808—eleven years before the appearance of Grimm's grammar.

From Schlegel derives not only the expression of the inner structure of languages or of comparative grammar, which is to serve as a basis for "the genealogy of languages," as comparative anatomy is for "higher natural history," [21] but also the idea of the essential advantage of inflecting languages (by inflection he means the inner alteration of the root) [22]—that they have arisen organically and "form an organic fabric." [23]

It is interesting to see that Bopp too seems to have taken individual conceptions of linguistic content from a vague knowledge of science. The foreword to the first impression of the *Vergleichende Grammatik* (1833 ff.) begins with the following words: "In this book I aim at a comparative description, comprising everything related, of the organism of the languages listed in the title, an investigation of their physical and mechanical laws and of the origin of the forms which characterize the grammatical relationships." What Bopp means by this distinction is well known; by physical laws he means what we nowadays call sound laws, and as mechanical laws he characterizes his theory of the relative weight of vowels and syllables. How does Bopp arrive at this remarkable concept of weight? Bopp assumes that the forms comprising one paradigm are of equal weight: where the ending is light, the stem becomes heavy. The weights of stem and ending are thus inversely proportional. It can probably no longer be proved, but it is most probable that here Bopp had in mind chemical theories of the beginning of the previous century. This is supported by the fact that he not only spoke—in the image widespread at that time—of an "anatomical dissection" of languages, but occasionally also of a "chemical decomposition of the body of language";[24] his concept of "weight" could also be derived from such chemical motions. The discovery of the relation between the atomic weight and specific heat of an element by Dulong and Petit came in 1819—the year of the English edition of his *Konjugationssystem*. The specific heats of elements are related in inverse proportion to their atom weights; the greater the atomic weights are, by so much are they smaller, and vice versa; in other words, the product of specific heat and atomic weight represents a constant quantity for all elements.[25] Bopp's paradigm of stem and syllable was such a "product."

However, at this point we must return once more to the beginnings of experimental phonetics itself, as founder of which Wolfgang von Kempelen has every right to be designated. Of his *Mechanismus der menschlichen Sprache* (1791) Ernst Brücke writes that in it we have been left a physiological phonology," in which certain things have certainly been supplemented and occasionally even improved, but which was so firmly based that it has formed and will form the most secure foundation for all further researches." This work is one of the best physiological books he had ever read, and he recommends it particularly "to the linguists who wish to make themselves familiar with the purely mechanical part of phonetics." [26]

This judgment of Brücke's is so important because it contains most concisely his own attitude, so deeply rooted in the mechanics of biology, to the problem of speech physiology, which through him via Scherer[27] has penetrated later experimental phonetics. That in the investigation of the movements of the speech-organs, too, neither the presuppositions of mechanics nor those of comparative experimental physiology are sufficient, since these movements must be guided according to principles which can only be grasped linguistically, if the speaker wishes to be understood—this thought is as far from Kempelen as from Brücke. And although Kempelen expressly states his dependence on Herder, and quotes long passages from Herder's dissertation "Über den Ursprung der Sprache" and "Ideen zur Philosophie der Geschichte der Menschheit," yet the idea that a phenomenon is completely explained only when it is explained mechanically is quite dominant with him. His dependence on Haller's physiological discoveries and investigations and on Newton's ideas of the mechanical explanation of all movement processes is too strong for him to be able to free himself from them under Herder's influence.

One thing must however be noted: he keeps himself free of the idea of an original language—not because the idea would have been impossible to accomplish for him, but simply because he considers it lacking sufficient support[28] and because this idea does too much violence to his knowledge of living languages. This reason was for him as a Viennese very cogent, because he himself spoke Hungarian and it was naturally easy for him, for example, to compare the numerals of Hungarian with those of the other European languages. For this reason he considers it more prudent to regard common words as "borrowed," and attempts by means of a botanical image to refute the image of the branches of one trunk: "As little as the apple-tree, oak, or lime have sprung from the pine-tree, just as little can all these 120 numerals be snipped out from 10 words of an original language." And he has yet another striking example—again biological—against the idea of an innate original language, from which all now extant languages have developed "like daughters of one mother":

> If it has really been implanted in man, how has he been able to lose it, or to alter it? Who can alter the circulation of his blood? Would this innate language have been less the work of the Creator, less permanent, than our pulse? If God has implanted a language in man's nature, if he has sent forth from his hand this creature destined for gregariousness, already fully equipped with the main requirement for this, then this innate gift would have to be reproduced with him, just like all the other bodily characteristics, like the cry of animals. A man who had been lost in the wilds in his childhood would, when he returned as an adult to humans again, speak it properly by himself without having to learn it first. The domestic cock still crows today as it crowed a thousand years ago, the white parrot still calls "cockatoo" as it did on the first day of its creation.[29]

His view is, consequently, that through an "unopposed custom, which has gradually become generally accepted and finally law," the first "inventors of language" agreed among themselves, and that now the task would be "to investigate the difference between the languages"—with regard to the words and the syntax. These are—if one considers with these the significant remark about records and the idea of the borrowing of words—thoroughly historical viewpoints, and to this extent he stands much closer to linguistics than experimental phonetics did decades after him. But finally he had a vague notion of a universal alphabet, which would have to be mechanically conceived.

Naturally it remains to be said that there was not as yet a comparative linguistics, which could have provided him with a guideline. However, it deserves to be noted that experimental phonetics has its roots in Kempelen—that is, before the development of the comparative linguistic method—and that not only the works on speech-physiology but also the phonetic investigations and problems of the whole 19th century were influenced by this idea.

Jacob Grimm had a very clear feeling for what was not in conformity with language in this apparently so exact research: "If one attributes purely physiological functions to the sounds, and bases on this an unproved and unprovable system of articulation (however much sagacity and tact is shown in this), for me at least the air becomes too rarefied and I am unable to live in it." [30]

Merkel, as one can imagine, criticized him for this very passage 26 years later;[31] but if one reads Rapp's "Versuch einer Physiologie der Sprache, nebst historischer Entwicklung der abendländischen Idiome nach physiologischen Grundsätzen," [32] which appeared four years before the third impression of the *Deutsche Grammatik*, one will understand that for Grimm "the air" could become here "too rarified." It is sufficient to read through the foreword, the introductory remarks, and the introduction—about twenty pages in all—to see how little Rapp noticed that through Rask, Bopp, and Grimm new tasks and methods appropriate to these had been created, which could no longer simply be forced into the bed of some type of speech-physiology—not to mention that the fact escaped him that physiology also had meanwhile found its own task through the comparative method, and no longer extended to the mechanical explanation of life-processes after the model of physics; but that meanwhile physics and chemistry began to become auxiliary sciences for it, which had to become subordinate to the task of comparative anatomy, comparative physiology, and comparative developmental theory.

One could object to Grimm—as Merkel did—that physiological considerations were far from his thoughts. "In conformity with this," Merkel writes, "he actually called his phonetic theory a letter theory and so generally in the concept of the sound did not get past that of the letter." However, Merkel forgets not only to mention that Grimm in the 3rd impression of the *Grammatik* did use the new expression "Lautlehre" [phonology] [33] for his letter-theory, while retaining the abovementioned word in the foreword, but that the view according to natural law—and certainly in a form more suitable to the object—is actually the achievement of Grimm. "And without mild surprise, indeed with secret laughter we read in these letters [in the letters of the brothers Grimm to Lachmann] that the expressions which today have long been familiar to us: 'Anlaut,' 'Auslaut,' 'Lautverschiebung,' 'starke and schwache Flexion' [initial position, final position, sound shift, strong and weak inflexion]; further, the expressions that derive from Kant and Wilhelm von Humboldt 'organisch' and 'unorganisch,' and then the synonymous fruitful expression 'Lautgesetz' [sound law] were at that time newly found and ventured hesitantly." [34]

One must take Grimm's statement literally: it is not the refusal to investigate *also* the positions and movemens of the speech-organs in the pronunciation of sounds which follows from it, but to attribute to the sounds "purely physiological functions" and to base "on this an unproved and unprovable system of articulation," which is here summarily refused by him. His criticism attacks the root of speech physiogy and at the same time that of experimental phonetics.

It is perhaps symptomatic that the compounds of the word "Laut" just quoted were created by Grimm himself and were anchored in linguistics, while the concept of phonetics formed, as it were, behind his back.

In Rapp we find in 1836 a distinction between the quality of the different sounds, the quantity, the modulation ("However, it is a particular type of modulation, which is discussed here. In many districts the whole indifferent speech-material is, one could say, sung to a certain melody; it is a certain cadence, a changing, rising and falling of the voice"),[35] and the difference between loud and soft ("energy of articulation"). One notices his concern for clear concepts, for which, however, he lacked the necessary empirical knowledge of the situation.

The turning away from texts to the spoken language was begun already in the first generations of modern linguists. The importance of Andreas Schmeller[36] lies—apart from his other learned works—not only in the scientific investigation (founded by him) of German dialects, whose phonology and grammar he was the first to make productive for the knowledge of the Old German situation also, but—far beyond the confines of dialect-study—in the evaluation of the *living* language, of the *living* dialect, which he obtained from the Bavarian vocabulary as well as from written sources. In 1821 his book *Die Mundarten Bayerns grammatisch dargestellt* appeared; and from 1827 to 1837, the first four volumes of his *Bayerisches Wörterbuch*, whose treatment of living dialects protected him from the error "from which historical grammar has only very gradually freed itself—that of the confusion of sound and letter" (E. Schröder).

Rudolf von Raumer, the Breslau-born son of the Erlangen mineralogist, Karl von Raumer, arrived in Munich for the summer term of 1836. Like Schmeller, the Raumers came originally from the Upper Palatinate, until they were transplanted to Anhalt by the Counter-Reformation. Raumer, who was then twenty-one years old, attracted by Schelling's philosophy, had personal contact in Munich with the Berliner Hans Ferdinand Massmann, who had been Professor of German Language and Literature there for seven years, and with Schmeller, whose dictionary was almost completed. A year later, Raumer (before he received his doctorate) published his book "Die Aspiration und Lautverschiebung," in which a sharp distinction was made (in Schmeller's sense) between the spoken and written language, and a series of insights expressed and demands made, that must be regarded as the "proliferation points" of the youngest branches of German linguistics. First, in this book of 1837, the problem of exceptions to Grimm's Law of the Sound-Shift is treated, which forty years later was to unite the Neogrammarians under Brugmann and Leskien and to give Wenker the stimulus for his questionnaire activities, which determined the development of German dialect-studies. So the movement of linguistics and dialectology to phonetic consideration and argumentation does not start in the adoption of Brücke's phonetics by Scherer as is sometimes asserted, but here already: "Here we will scarcely reach our goal," Raumer writes with reference to the transition of aspirated stops to voiced stops, "if we content ourselves with having found certain letters in certain dialects in the place of other letters in other dialects. We must go into the nature of the sounds designated by these letters, in order to see how the one can develop from the other." [37]

Etymology, then, is not to be replaced by phonetics, nor is "the phonetician replaced by the historian" (F. Wrede), and here nothing is said of a twofold division into linguistics and phonetics, but of co-operation of the two, which indeed only temporarily took place after Wilhelm Scherer's "Habilitation" in Vienna in 1864, where, eight years before, Ernst Wilhelm Brücke had published his principles of the *Physiologie und Systematik der Sprachlaute*. Four years later (in 1868) Scherer published his work *Zur Geschichte der deutschen Sprache*, which applied modern phonetics (in Raumer's sense) to Germanic phonology.

If one can therefore regard this demand of the young Rudolf von Raumer as a root of the "phonetic epoch," yet there are in this, his first work, also suggestions to which only few linguists of the last century returned. He is the first to go into the "fluctuation," the "variability" of sounds *expressis verbis*, into "the whole series of intermediate stages" between the "extreme limits," which are expressed by special letter symbols.

Two factors are drawn into the field of vision with him for the first time: discrete phonetic symbols and the continuum of producible sounds. The *connection* between these two factors and their proximity to statistics Raumer did not yet see, although the book by Johann Peter Süssmilch was available since 1741 [38] and the investigations of Karl Friedrich Gauss on the theory of the errors of observation [39] since 1823, and F. B. Wilhelm von Hermann had been teaching for four years previously as Professor of Economics at the University of Munich, and his *Staatswirtschaftliche Untersuchungen* had appeared in 1832; that is, five years before Raumer's first work: three works in which solutions for the problems begun by Raumer could have largely been found. "All these questions," Raumer ended his book, "will be resolved only by 'a comparison of everything comparable' with evidence." [40] However at that time he could not yet perceive by what methods this "comparison of everything comparable" with reference to language—and for him that meant the spoken language—could be brought about.

Rudolf von Raumer did not content himself with this general demand. Twenty years later he wrote an open letter in the 4th volume of the periodical *Die deutschen Mundarten* to its editor Georg Karl Frommann, director of the archive and library at the Germanic Museum in Nuremburg. [41] He writes in the letter:

> It seems to me, in the present stage of linguistic research, to be of quite particular importance to grasp language in its most individual phenomena with the greatest possible precision and exactitude, and to this end no

branch of linguistics offers the means to such a degree as the observation of living dialects. . . . We cannot of course think of following all the linguistic idiosyncrasies of all individuals. However, it is of the greatest interest to become familiar with the real speech of different individuals from one and the same region with diplomatic exactitude at least *in selected cases* . . . But even where the aim of the author . . . is the communication of a specimen of language for scientific ends, it is usually not the language of any particular person which is communicated but *what is common to the regional or local dialect after the removal of all that is individual*. Contrary to this then would be what I mean, that we should observe quite definite persons and reduce their speech to writing as faithfully as possible. Such a communication would be related to the method usual up to now as a portrait is to a historical painting. And the portrait too would be understood for our purposes not in the idealizing manner of the artist, but in the strictly reflecting manner of the daguerreotype.[42] If we had a machine that would collect what is spoken just as faithfully and fix it to the paper as a daguerreotype does with what is seen, then its achievements would correspond to what I wish. Since, however, we do not have such a machine, we must attempt to approach, at least to some degree, what it would offer us. Understood unconditionally, the matter has insurmountable difficulties. But for this very reason one can only speak of a greater or lesser approximation. I say advisedly "of a greater or lesser approximation." For even with the sacrifice of all refinements difficult to establish, such an attempt, even rough hewn, would already be very profitable. Then, first, we would be sure that we had before us the sentence structure of the speaker without any admixture from the recorder. This is a matter of much greater importance than many think. For we would receive a quite false impression precisely of the sentence structure of the really spoken dialects, if we kept to many so-called dialect specimens. Secondly, such a "portrait-making" record would communicate to us under all circumstances the grammatical forms really used by the speaker. In this regard also, not a few of the usual specimens of dialect leave us very much in uncertainty. At times, the writer mixes in forms which probably would sound thus (by analogy with the whole dialect) as the writer gives them, if they really did exist in this dialect. But they do not appear in it at all. At other times, the writer regulates the actually existing forms in such a way that *he removes as far as possible all fluctuations* and carries through in his communications *only one* of the forms used, which appears to him to correspond most to the whole analogy of the dialect. This all disappears, as soon as we have a mirror-image of what a certain man has actually spoken. Finally, the sounds of the speaker too should be reproduced as faithfully as possible. Here we enter an area in which only relative exactitude is attainable. But even with this, quite a lot would be accomplished. Particularly, the *phonetic fluctuation of many word-forms* could be represented to a considerable extent even with the means of phonetic transcription used up to now.[43]

If we compare to this program the preference for the problem of sounds and their changes at the time of the Neogrammarians, we can admire the singlemindedness with which Georg Wenker pursued his problems and brought them towards a solution. It is interesting to see, from the point of view of the history of our science, to what extent Wenker, through this action of his, could put almost the whole of German dialect study under the spell of his question.

That appears even more clearly from the following elaborations by Raumer in the same letter:

> The content of such [that is, fixed in the manner of the daguerreotype] communications could be very different. If it consisted of a fairy tale or some other narrative, a few words should be added on the station, age, and type of the narrator. . . . It should be indicated each time, however, or according to certain headings, to which class the person belonged *with whom the speaker was speaking*. In this way we could choose several persons from one and the same place, of whom we would record a series of isolated utterances in the manner indicated. If we chose three or four people who retained the idiosyncrasies of the dialect in particular clarity (although, as would appear, *by no means in absolute equality*) and a further three or four whose speech *approximated gradually more and more to Standard German*, a series of such recordings would give a clear picture of the *actual spoken language of the inhabitants of the place*, such as we cannot gain from other types of representations. Besides, it would also be of interest to listen to *people from other areas whose original dialect had gradually approximated that of their new home*, in their idioms . . . The best thing would be the communication of complete conversations, as they are really carried on between different people. But however helpful such a communication would be, we must be reminded most urgently of our real aim: to keep our eyes firmly fixed on *historical reality* in the strictest sense of the word. Do not think, my esteemd friend, that I consider this type of communication easy. I consider it rather extraordinarily difficult. But I believe also that I recognize in several of your esteemed colleagues the right people to fulfil my wishes. Should you decide to do this, you would certainly be rendering a very profitable service and not only to me, but to the *whole of linguistic research*.[44]

They did not decide to do it. And the subsequent generations did not do it either, not even when for decades—through the invention of the phonograph, the record, and finally the tape—all the technical possibilities for it were available.

> In 1843, two theories of sound were almost simultaneously advanced, which feuded violently for a fairly long time, until finally an astounding event took place. One theory derived from August Seebeck and asserted

that a sound depended on the laws of motion of the vibrating point; therefore, chronographically speaking, on the shape of the symbolic sound-curve: every such periodic the ear receives as a particular, specific sound. He succeeded in establishing and confirming his theory by experiments of various kinds. But the other theory, which was advanced by Georg Simon Ohm, could let the facts speak for themselves, too.[45] It had occasionally been observed earlier, as by the famous musician and music-theoretician Rameau, that when a note sounds, besides the primary note still higher notes can be heard; namely, the octave of the primary note, its twelfth, the double octave, its third, etc.; in short, those notes whose frequency is double, three times, four times, etc. the key note; these tones are called harmonic overtones of the key-note, but altogether partial tones. Taking C as the key-note, for example, gives the system [shown in Figure 1] of partial tones with their numbers (that is, relative frequencies), where those notes which do not correspond exactly to the tones concerned are designated by an asterisk [Figure 1].

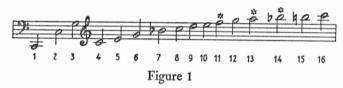

Figure 1

As can be seen, more than half the notes belong to the C major triad, the others representing very different types of notes, in part quite foreign to the C major scale. By systematic attention and practice the point can be reached where we hear these overtones with the unaided ear more and more clearly, and the detection is even more successful if we aid the ear by suitable resonators. Here we see that the overtones are of varying strength according to the way in which the base-tone is produced; on the violin, they have different relative intensities than on the piano; on the flute, other than on the trumpet; and, as far as the human voice is concerned, a different intensity for each vowel. Perhaps the most interesting thing in this connection—and that because of its negative behavior—is that the tuning-fork, especially when mounted on a resonance box, when struck with moderate force has no overtones at all; and if its oscillations are registered chronographically, we find a typical sine-curve. Ohm then advances the theory that only an oscillation of sine-character excites in the ear the sensation of a simple and single note; every other periodicity is analyzed by the ear into a series of simple notes, each of them likewise of pure sine-character. With some effort and practice one can separate out these simple notes, but one usually merges them into a new unity; and this merging produces, according to the strength of the separate partial tones, the specific sound of the whole tone. It should moreover be noted that the theory in this compact form was only advanced after Ohm by Helmholtz, who then built up his whole system of sounds with this theory as its essential foundation.

Thus, to sum up: according to Ohm the ear perceives only sine-oscillations as simple notes, all others as compound ones, and in the latter case it blends the notes according to their relative intensities into a specific sound; according to Seebeck the ear perceives every oscillation as a matter of course, as a unified note, whatever law of motion it may follow, and according to the law of motion as a note of a different sound. Which of the two theories is correct? The dispute did not last long, for it very soon appeared that the two theories do not exclude each other, that on the contrary they are only different ways of expressing one and the same theory; and the proof of this had been given already decades before by mathematical analysis and synthesis: it only required appropriate application to the problem under discussion. It is here a question of a topic which is so interesting, not only for the problem concerning us, but quite generally for the nature of human intelligence and perception in their relations to each other, that we must here consider it rather more closely.

Since the middle of the 18th century, the question had been considered: what would be the simplest periodic function?—function in the sense of a quantity which necessarily alters when a quantity, on which it depends, is arbitrarily altered. For example, temperature is a periodic function of time; it fluctuates gradually up and down (on the average from many observations) between a minimum in the early morning hours and a maximum in the early afternoon. ... It now became apparent that among all the possible periodic functions there is one particular one, which represents the basic type, so to speak; namely, the sine-function—for the reason that one can construct all the other functions from it. For example, take the lightly-drawn sine-curves in [Figure 2]—for the strongest of which only a single wave-crest is drawn, for the second a trough, a crest and a trough, etc.—and sum them up at each point. The more such sine-curves of suitable height one employs, so much more exactly one obtains the heavily-drawn line, which is then not a sine-

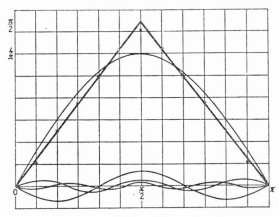

Figure 2

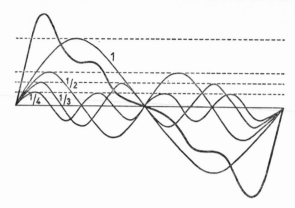

Figure 3

curve, but the quite variant figure of a symmetrical roof with straight slopes. Such a roof-line cannot be directly represented mathematically in a unified form (for in the first part its course is quite different than in the second); however, it can be expressed as a series of harmonic sine-members, each member being provided with a definite sign (crest or trough in front) and a definite amplitude-factor (height and depth of crest and trough). In the above example the formula is

$$f(x) = (4/\pi) \, [\sin x - (1/9) \sin 3x + (1/25) \sin 5x - \ldots]$$

and the more members of this series that are considered—the more sine-curves that are drawn and then totalled—so much the more exactly does the construction of the roof-line occur. Or consider [Figure 3]: here we have a very slow sine-curve of great amplitude, one of double the speed but only half the amplitude, and a third with one-third and a fourth with a quarter the amplitude; the result is the strong curve, which again is of quite a different type. It rises first to the peak, then falls in several undulations to the lowest point, then returns in a regular sweep to the normal level. These two examples will suffice to show what enormous variety there is here in the construction of curves, and all this with very restricted building materials, namely, only sine-curves. Instead of saying that this periodic curve has this or that form, one can say (and indeed in a more exact and more informative way) that it is built up from sine-oscillations of such and such amplitudes. This whole mathematical theory was brought to completion by Fourier and the sine-series have since then been called after him "Fourier series," and the whole analysis is designated "Fourier analysis." Now we comprehend how the theories of Seebeck and Ohm differ: Seebeck takes the curves—in this case the chronographic image of the periodic sound-movement—as a whole; Ohm, on the other hand, analyzes them into their Fourier elements; the one is thus, primarily at least, just as correct as the other. We can only hope to reach a decision if we turn away from the mathematical (or conceptual) side of the question to its physiological (or sensual) side; in other words, if we consult the ear. But the answer is already given: the ear proceeds, when left to its own devices, like Seebeck—it functions synthetically; but we can train it to proceed like Ohm—that is, to work analytically....

... Whether with the unaided ear or with the help of resonators, one can after some practice isolate from a sound a great number of the overtones, most easily with the piano and the human voice. As far as the latter is concerned, when the vowel u is sung, one hears besides the fundamental almost only the octave; with o, one hears particularly clearly the twelfth, the second overtone; with a, some middle overtones; with i, some fairly high ones are particularly strongly represented. How susceptible the method is can be seen by the fact that if the same vowel is sung once low, again high, the strength of the overtones varies noticeably and in fact corresponds to a variation in the sound....

In analysis the ear is naturally of limited efficiency, and incapable of competing with what a mechanical or (even better) photographic representation of the sound-curves gives with regard to details. All the more important instrumental and vocal sounds have in this way been made visible to the eye, particularly by Helmholtz and his disciples, and present an astounding picture of their diversity.[46]

The discussion of the history of vowel analysis after Helmholtz would lead us too far afield at this point. We only mention briefly that for whole decades it was a question of sung, isolated vowels, which are associated with particular circumstances inasmuch as they are based on periodic or almost periodic movements of air-molecules—practically in the same way as with instrumental sounds. We refer the reader to the descriptions by Kalähne,[47] Crandall,[48] and Fletcher,[49] and to the eighth volume of Geiger and Scheel's[50] *Handbuch der Physik*, edited by Trendelenburg.

With spoken sounds, however, it is primarily a matter of nonperiodic processes, even where isolated quasi-periodic processes occasionally occur—for example, with long vowels. It is part of their nature that they have a beginning and an end, which—since a sound has a semantic function only in the context of the word and sentence and is only through this a *speech*-sound—cannot be ignored; whereas this division is allowed with instrumental sounds and sung vowels from physical points of view.

This appears to be a purely empirical situation, and the endeavors of experimental phonetics in earlier years have shown that it is regularly so interpreted. It was argued as follows: with instrumental sounds it is a matter of purely periodic functions, so one can without real error interpret sung sounds as such. In the case of spoken vowels and voiced sounds, the periodicity recedes further into the background, and with voiceless sounds there is no periodicity at all; instrumental sounds remain the model for the idea of the construction and structure of sounds; with consonantal voiceless sounds it was a matter of complicated exceptional cases. That this argumentation is not in order can be easily proved; here we will only draw attention to the epistemological side of the situation, so far as it is characteristic of the historical development of phonetics.

First of all one must naturally say: a method cannot be characterized as a method of sound analysis, if only some sounds—namely, the voiced ones—are subject to it, but other sounds are not. The ideas of Fourier analysis—periodicity and the construction of sounds—have physical character in the strictest sense of the word. They conform to the Galilean concept of the experiment, of the construction of a phenomenon, and to the idea of the union of mathematics and experience confirmed in the experiment. A sound becomes a natural object in the sense of exact science only when it is defined with the help of Fourier analysis through its mathematically comprehensible and (for example, by resonators) perceptible partial tones. Without this mathematical construction it is, platonically speaking, not removed from the flight of phenomena—without this, not of interest to the physicist either.

But what of the spoken sound? A speech sound exists—that is, it is linguistically definable—not when it can be physically constructed, but when it is recognized as a speech sound in the service of communication. It is thus not only thrown back on perception in the other sense than sound, even above or below the limits of hearing, to speak of which still remains physically significant (that is, when perception no longer occurs and according to experience cannot occur); but it is at the same time related to a traditional system of means of communication. In a scientific sense it is defined when it is understood in the context of a traditional structure. If it were otherwise, there would have been no linguistics until today; for the physical construction of the sounds is still known sufficiently of but a few languages.

In 1870 and 1876, independently of each other—by the Russian inaugural lecture of Jan Baudouin de Courtenay (which became known in Western Europe only through Roman Jakobson and Mathesius) and by the dissertation of J. Winteler on the "Kerenzer Mundart des Kantons Glarus" (which also became famous only lately, but rapidly)—two essentially connected distinctions were drawn, which became a *sine qua non* of all future quantitative linguistics: by Baudouin, the distinction between traditional linguistic signals and their (physical) realization in the act of speech; and by Winteler, the distinction between phonetic differences with and without the function of semantic differentiation.

In 1876 Georg Wenker sent out his first questionnaire, by which he transformed the Neogrammarians' thesis of the universality of sound-laws into a question that he could and would test on Rhenish dialects by linguistic methods. The fact that this proof did not succeed in the way he expected made him the founder of the "Deutscher Sprach-atlas," which has become a model for the world's language atlases: he sent to 40,000 villages with schools in the German Reich of that time the sentences named after him, to be translated into each respective "local dialect."

Raumer's idea of the variation of "the language actually spoken by the inhabitants of the place" had to be abandoned in favor of a uniform local dialect, and his demand that "complete conversations" be recorded "as they are really conducted between different people" had to be given up in favor of the translation of Wenker's forty sentences. Neither were the distinctions of Baudouin de Courtenay and Winteler taken into consideration. The "constituent factors" of language: speech-melody, accent, relations of sound-duration, pauses, separation of syllables, etc.—of such decisive importance for the differences between languages—could not be represented by symbols at all, or only inadequately.

Without being able to demonstrate by observation Raumer's "inevitable . . . phonetic fluctuation," Hermann Paul (then Professor of Germanic Studies in Freiburg) subjected this fluctuation to penetrating theoretical discussions in 1880. He distinguished "sound-image" and "feeling of movement" from "variation of pronunciation," which was controllable only to a limited extent. He also separated the theoretical symmetry of these variations from the "actually occurring deviations," which were not distributed symmetrically "according to number and quantity," and he based his Neogrammarian theory of sound-change on these ideas.[51]

But already two years before this, Raumer's wish had been fulfilled by Edison's invention of the phonograph. At the same time it became possible to test Paul's theoretical considerations on the object—and now indeed really "according to number and quantity"—and to make these ideas of service to linguistic research.

However, the way there was long. Immediately after Edison's invention, Jenkin and Ewing in England, and in the 1880's Lahr and Hermann in Germany, succeeded in transforming the grooves of the phonograph or record into visible and measurable curves. But it was recognized that these technical possibilities could gain significance for linguistics only after it was understood that the value of the new type of record did not lie in the "objectivization" of hearing and of understanding and in the objective measurement of speech sounds.

In 1880 the phonograph was at least considered for phonetic investigations by F. Techmer—quoting the anatomist Cuvier, in search of a "palaeontology of language," and in 1883 by Carl Stumpf for psychological and musicological investigations. In 1887 the German-American Ernst Berliner invented the moldable record, which could thus be multiplied at will, and—in contrast to Edison's deep grooving—the shallow grooving named after him. In 1889 the Parisian anthropologist Azoulay established the first phonogram-archive of languages, which he was able to record during the Paris World Exhibition, and published the first catalogue.

In 1889 two doctors demanded that the record should be made of service to science: the Giessen psychiatrist Robert Sommer who, in his *Lehrbuch der psychopathologischen Untersuchungsmethoden* [Handbook of Psychopathological Research Methods], recommended the recording of "complete dialogues between doctor and patient," and the Viennese physiologist Sigmund Exner, who instigated the founding of a "phonogram-archive" of the Imperial Academy of Sciences. The special thing about this move was that Exner advanced this proposal together with representatives of other disciplines, who here for the first time united in a group, as the case required, that showed the central position of language. Those advancing the proposal were, besides Sigmund Exner, the physicist Franz Exner, the classical philologist Wilhelm von Hartel, the Germanist Richard Heinzel, and the Slavist Vatroslav Jagić. After the acceptance of the proposal, a commission of members of both classes was appointed, which chose Sigmund Exner as its chairman—perhaps because he was the initiator, but perhaps also because physiology then occupied a position of precedence.

The interests of the turn of the century met the Viennese endeavors half-way. The tendency of linguistics towards the natural sciences, the investigations of Wenker and Wrede, revealed their first fruits: the demand for a systematic recording of the German dialects by Friedrich Kluge, Otto Behaghel, Karl Müller, and Kekule von Stradonitz; the founding of the "Allgemeiner Deutscher Sprachverein"; and finally the general desire for a"Reichsamt für deutsche Sprachwissenschaft," for a "Reichsanstalt für deutsche Sprachforschung", for an "Akademie der deutschen Sprache" to which these researchers had given public expression in 1900, 1901, and 1903. In 1904 the Jena scholar in Indo-European philology, Berthold Delbrück expressed the expectation that with the help of the record "phenomena of sound-change might be discovered, which under certain circumstances extend over generations."

Some ten years after the publication of the first impression of Paul's *Prinzipien* there appeared in Leipzig Hans Georg Conon von der Gabelentz's work "Die Sprachwissenschaft." [52] With all the clarity one could desire, the author—about 25 years before Saussure,[53] who is today regarded as the founder of modern linguistics—distinguished from the then generally practiced "genealogical historical linguistic research" (as whose foundations he named Grimm's *Deutsche Grammatik* and Bopp's *Vergleichende Grammatik*) [54] "general" linguistics and "individual language" research.[55] And in this work he gave the latter precedence over the other two!

> I will not be misunderstood if I place in sharp contrast the viewpoints of the research on individual languages and that on the history of language. I certainly recognize the equality of both. I attempt to show how both must finally be interwoven. But for this reason I would rather see the threads kept apart at first, than matted together [Foreword V].
>
> ... One is only too ready to imagine that one knows why something is now, if one knows how it was earlier, and knows the relevant laws of sound change. However, that is only true so far as these laws alone determine the fates of words and word forms. For example, if I know that Latin *f* becomes *h* in Spanish; *li* before vowels, *j* (pronounced χ); and the ending of the second declension, *o* in the singular and *os* in the plural; then it is explicable to me how *filius* must have become *hijo*. Granted then that each word and each form of the Spanish language were genetically derived in this way: would the Spanish language be explained by this? Certainly not, for the language is just as little a collection of words and forms, as the organic body is a collection of members and organs. Both are *(relatively) complete systems* in *every phase* of their lives, only dependent on themselves. ... It is not the egg, caterpillar, and pupa which explain the flight of the butterfly, but the body of the butterfly itself.[56]
>
> ... The task of research on individual languages is thus to comprehend a language solely as it lives in the mind of the nation speaking it. This nation manipulates its language without looking back to its prehistory or sideways at its dialects or foreign relatives; all the factors which deter-

mine the correct manipulation of the language lie exclusively within this language itself, and thus demand to be comprehended from it.[57] ... The language laws form an organic system among themselves.[58]

... One speaks of the *organism of language* with complete justification or at least without harm, as long as one bears in mind that language is not an independent being, but an ability and function of the human body and mind. If one recognizes this, then quite different affinities will emerge—genuine, genetically-based ones, not mere analogies. Religions, law, ethics, in short the whole cultural life of the nations is determined by the same forces as their languages; they can be determined by no other forces.[59]

Von der Gabelentz recommends connecting every description of the grammar of an individual language with an "ideal grammar," whereby he demands "suitability of method and nothing more." [60]

Quite important is his incidental remark that language is "simultaneously object and means of description . . . in grammar," therefore simultaneously fact and principle—a problem, which from the psychological point of view was not solved until a generation later by the psychology of thought, and which for the principles of linguistics and dialectology is not even approaching exhaustive treatment, even today.

From the difference between the unidimensionality of linguistic description—Saussure's principle of linearity on the so-called "syntagmatic plane"—and the bidimensionality of the faculty of speech on the so-called "paradigmatic plane," which moves towards tabular description, he derives his distinction between analytic and synthetic grammar, to which he prefixes the theory of the inventory of sounds. "By this theory I understand the systematic enumeration and description of the sounds (that is, of the individual languages or dialects) and the indication of what positions and what combinations they can appear in; the description of accents will follow." [61] These are investigations which—following Saussure—were first taken up systematically by phonemics and structuralism, and have largely still to be carried out for the German dialects. Saussure's importance will not be reduced by these references to the work of Gabelentz. But a higher estimation of this original researcher and thinker would have been useful not only to German dialectology.

In the years between 1906 and 1911, when Ferdinand de Saussure was giving his lectures—a time of overwhelmingly historically-oriented research—he not only distinguished between "diachronic linguistics" and "synchronic linguistics," which is concerned with the relationships "which exist between contemporaneous members of a system," [62] but also (like Baudouin de Courtenay before him) between this "science of language" and a "science of speaking" that he subordinated to the former—at that time a completely new idea, whose importance has only been recognized during the present period. He thereby directed the attention of his epoch again to the description of perceptible phenomena and thus prepared the path which Raumer had already indicated: to record conversations between people on sound carriers and on these "to establish the basic principles of each idiosynchronic system, the formative basic factors of each state of language." [63]

After the development of the low-frequency amplifier, it became possible to make recordings electrically. Since with this procedure it was no longer necessary to speak into a trumpet, real conversations between several people—even without their knowing—could now be recorded by means of a microphone. The wish of Raumer—Utopian in his times—and the demand of Robert Sommer had become actually realizable.

Further important possibilities were offered to research by the "Triergon process," by which the kinetophone, invented by Edison in 1893, was developed further into the modern sound-film and the X-ray-film. However, all these opportunities of relating with each other the audible word and the measurable sound-pressure curve and later also the motion-curve obtained from the reel of film, were at first not utilized.

In 1876 J. Winteler wrote in the *Kerenzer Mundart*:

As far as the division of the articulation possibilities (varied according to place) within the two articulation pairs or subdivisions goes, this is extremely fine. Empirical language moves in innumerable transitions, and the cases already described in sound-physiology and linguistics describe only the more profound differences: numerous intermediate forms lie between these boundaries, and it is precisely these which contain the key to the understanding of the historical sound-development which is established only according to the most obvious differences. This diversity of articulation nuances within such simple basic features has its explanation not only in the possibility of variation in the wave lines (peculiar to the articulation of the tongue), but also in subjective factors. Since a significant number of muscular fibers and nerves of the sense of touch are involved in every articulation, nuances are possible within an articulation not further to be differentiated, according to whether or not the speaker prefers to place the primary emphasis of articulation on the activity of one or another of the muscular fibers and on the perception of this or that aspect of their effects. If we now consider that such fine local differentiation is true of sound-forming as well as of sound-modifying articulation, and if we consider also that several such finely nuanced articulations may appear in a sound-modifying capacity in the most diverse combinations, and that finally the variation in articulation is not merely a local one but also occurs with reference to degree, form, and duration, then we will gain an idea of the *possibilities of variation approaching infinity in those types of sounds which from script we are accustomed to regard as constant and as very limited in number*. However, it is those fine nuances, and not these few abstractions, in which living language moves and on which there hangs, like a scent, precisely what has been preserved from the evolution of language in the consciousness of the speaker of language (with reference to the phonetic aspect). Despite the difficulty of the question, we therefore cannot dismiss from our minds the task of approaching those subtleties as closely as possible. Naturally this is possible in no other way than by observation of the living vernacular, which has developed as far as possible uninterruptedly and naively.[64]

When Trubetzkoy, in his *Principles of Phonology*, pointed out the differences between the views of Saussure and Baudouin de Courtenay,[65] it was for him primarily a matter of showing that the differentiation between language structure and speech act was only hinted at by Saussure, whereas Baudouin de Courtenay also demanded two corresponding types of description, whose differentiation Saussure did not expressly require. It corresponded to the early conception of the Prague School of phonology to place the stress primarily on this aspect of the difference between the Swiss and the Polish researchers. As a result of this conception, they saw phonology at the level of grammar as part of the humanities and social sciences and demanded its consistent separation from phonetics in the sense of the natural sciences. Independently of the Prague phonology, phonemetrics has since 1932 stressed the dependence of the phonetic point of view—physiological as well as physical—on the linguistic one,[66] whereby it had necessarily to return to the idea of Baudouin de Courtenay that the object is subordinate to the viewpoint from which it is scientifically approached. In the meantime, the dependence of quantitative phonetics on phonemics—and at the same time its connection with it—has appeared much more clearly than in the years when Trubetzkoy still saw himself obliged to protect his field from the claims and encroachments of an experimental phonetics that was regarded as an independent natural science. Discussion of the relationship of phonemics to phonetics[67] has made us more aware that it is not—as Karl Bühler has falsely asserted to the Prague phonologists—here a matter of two different objects: language structure *and* speech act, which then on their part require two independent sciences (rather as Germanistics and Romanistics are required by the existence of the Germanic and Romance languages).[68] Instead, it is a matter of two aspects from which one and the same object can and must be treated. To have perceived this seems to us to be the decisive merit of Baudouin de Courtenay as against Saussure and his pupils.

In this connection it is irrelevant that Baudouin de Courtenay did not fully perceive the relation of linguistic research on the one hand to psychological, physical, and physiological observation on the other and that he did not quite realize the dependence of these observations on the linguistic model.

These problems have come further into the foreground since it has been more clearly perceived that experimental phonetics was never concerned with sounds on the physiological or acoustical level alone. Experimental phonetics also dealt with problems of accent, and these were necessarily related to syllable and word. Its endeavors with regard to speech melody, from their beginning, have turned towards the sentence melody. However, linguists and phoneticians have become more conscious of these questions only since the concept of suprasegmental structures has been formulated.

This means that the psychological, physical, and physiological aspects of phonetics are not justified by the existence of other speech sounds than sentences, words, and syllables, but by the fact that phonetics is a part of traditional linguistics, to which phonology has also belonged since the appearance of the Prague phonologists. Phonetics is indeed one of the aspects of linguistics, and it is subordinated to the aims of the latter. More precisely one could say that within itself it contains a system of different aspects, such as the psychological, physical, and physiological levels.

The psychological level has two components; namely, the physical or acoustic, and the physiological or articulatory ones.[69] The central problem of phonetics concerns the dependence of this level on structures transmitted in geographical, historical, and social terms. The relationship of phonetics to linguistics is based on a system of scientific questioning and investigating. These methods create their scientific objects as a system of relations, in the sense that these objects can only be defined by the research methods; they cannot be separated. This is the important result of the epistemological work of the late nineteenth and early twentieth centuries.[70]

In the years that have passed since the First International Congress of

Phonetic Sciences of 1932, this central problem of phonetics has become to an increasing extent the central problem also of the linguistic treatment of living languages and dialects.

It is here, both in the auditory and acoustic sectors of phonetics and in the intentional and physiological sectors (earlier distinguished as genemmatic and genetic phonetics), a question of the measurement of curves which represent events taking place in time, and therefore are represented in coordinate systems (with ordinates of changing value) whose abscissae represent the time. These curves are therefore continuums. But a continuum "never contains an impressed measure. . . . A chain . . . has by its nature a system of measurement, one can number the links. Not so with a continuous thread, which has no splits, as each split would interrupt the continuity. If one wishes to measure it, one must, like the tailors of old, place a ruler beside it and transfer its division." [71] For the physical measurement of time, some periodic natural events (such as the alternation of day and night) have been used for this "as far back as historical tradition reaches" (von Laue). In phonetics, however, it is not time but the natural correlates of traditional languages that are to be measured. And for this we then need, apart from the possibility of measuring time, an articulated structure, which we can lay beside the continuous curves, with which we can correlate the continuous acoustic or physiological curves.

This "articulated structure" is the perceptible and intendable, or perceived and intended, succession of realizations of a finite inventory of traditional and linguistically definable discrete occurrences of sounds, syllables, and words, with which curves of physical and physiological events can be coordinated, which without this succession of discrete and thus countable contents of perception or intention cannot be suitably articulated and partitioned off from each other in measurable form. Scripture[72] in 1930 failed to see this problem, just as Stutterheim[73] failed almost 30 years later.

The problem of the realization of such structures involves—in the auditory and physical sector as in the intentional and physiological—the problem of fluctuation; it was already hinted at in 1837 by Rudolf von Raumer,[74] and clearly recognized by Hermann Paul[75] about 80 years ago. Phonometrics[76] has shown that it can only be mastered scientifically with the help of the statistics of variation, which the Belgian astronomer and statistician L. A. Quetelet[77] in 1871 utilized for the mastery of the problem of biological distributions.

The Prague phonologists originally also spoke of sound-conceptions in order to distinguish the phoneme from its realization in the speech act on the one hand and from writing on the other hand, and originally they misplaced the boundary between phonemics and phonetics, and did not sufficiently stress the connection between the act of speaking and that of hearing; but their historical merit remains that of having investigated the new line of research first seen by Baudouin de Courtenay, von der Gabelentz, and Saussure. The first important step in this direction was to distinguish between semantic oppositions and other auditory or articulatory differences, and to point out the importance of phonemic opposition as the certainly predominant form of features that differentiate meaning.

It was noticed only relatively late by phonometrics that this discovery of the phonologists must be of decisive importance for the whole field of phonetics, and that this discovery affected phonetics in a twofold way: in the necessity of a variation-statistics treatment of the realization of both members of an opposition,[78] and on the other hand in the mathematical treatment of the realization of the opposition itself.[79]

It seems not to have been noticed before that Jacob Grimm was the first to express the idea that, with the relationship of the realization of opposition members, it was a matter of quotients: "The long vowel has double the measure of the short vowel," he says in the first book (entitled "Lautlehre") of the 3rd impression of his *Deutsche Grammatik.*[80] After him, Brücke, Krauter, and Viëtor have mentioned such quotients. However, they never carried through in this context the distinction between phonemic and phonetic considerations, and thus the consequences could never be drawn from these reflections, which could have been of advantage for the characterization, differentiation, and comparison of languages and dialects.

As long as experimental phonetics held the opinion that its task consisted in the causal analysis of the members of a universal alphabet, it could only link up with the historical and geographical tasks of linguistics in the few cases where it investigated sounds which only appear in a restricted group of languages. For this reason "tongue-click" sounds were from the beginning a favorite subject for the investigations of experimental phonetics, and the merits which it has acquired in this field should not be contested. However, real links with the linguistic formulation of questions could be acquired only after quantitative phonetics too had found a definite relation to phonemics, and the newly developed twofold significance of phonemics for the investigation of the realization—namely, that of phoneme *and* of opposition—was recognized.

III. Methodological
Foundations of Phonometrics

One of the conditions for phonometric investigations of languages is that we be able to *distinguish* segments. This is the basis for counting[1] segments and for all forms of measuring the characteristics of individual segments. Among the further conditions for phonometrics is that we be able to *distribute* these segments into a finite number of different classes.

In particular, a judgment as to the frequency with which certain segments appear can only be passed if these segments can be counted and grouped into classes that allow us to decide on principle as to whether each element belongs or does not belong to one of these classes.

With the positing of the concepts of *distinguishing* and *distributing* segments we subject the whole material to a *mathematical* method of consideration. When we turn towards only this one aspect of the structure of linguistic segments and disregard especially what linguistic functions these segments of conversation or speech have to perform, what they represent in terms of physics, and how they are physiologically produced, they become a problem of mathematics. One of the basic concepts of this science is founded precisely on the "logical distinction of things" and on the "logical judgment as to inclusion or exclusion." In these two criteria, mathematics sees the essential conditions of the so-called "set-theory." [2]

1. TWO OBJECTIONS TO THE DISTINCTION AND DISTRIBUTION OF LINGUISTIC SEGMENTS

Against this claim of phonometrics and of the whole of linguistics—namely to operate, in the framework of its investigations, with speech-sounds which are distinguished and distributed into classes—misgivings were voiced on the part of experimental phonetics, especially by E. W. Scripture. In what follows, we shall attempt to explain these arguments, in order subsequently to test their tenability.

Essentially, it is here a matter of two diverse objections, which can be characterized as objections at the *syntagmatic* and at the *paradigmatic* levels.

Syntagmatic Objection

In the "genetic" as well as in the "genemmic" fields of experimental phonetics, natural movements can be registered; that is, represented as visible and measurable curves: the movements of the speech organs of the speaker (of the mouth and nose, larynx, and thorax) on the one hand, the movements of the air molecules which stimulate the ear of the hearer on the other hand. In both cases, the continuity of these curves represents not only the inertia of the registering apparatus, but at the same time the continuity of these movements themselves and with them the inertia of the *anatomically* defined organs of speech and of the *chemically* defined molecules of air which transmit the sound waves.

Observations such as mathematical analyses of these speech-physiological or speech-physical curves indeed enable us to define the beginning or end of a speech phase before and after speech pauses with sufficient accuracy and at least to say with certainty that certain horizontal phases of these curves represent speech pauses; and the phases following or preceding these, speech-processes. However, within these phases one can distinguish with certainty only phases of faster alteration from phases of slower alteration, and this both optically and mathematically. This can be shown by Fourier analysis of extracted portions of curves, whose "similarity" to preceding and subsequent portions must here be regarded as mathematical equality, in order to satisfy Fourier's requirement for the periodicity of the processes analyzed.

However, neither the observation of these curves nor their mathematical analysis allows us to see the transition from their continuity to a succession of definable segments, which establish the existence of the discrete speech-sounds which we believe we speak and hear. Such a conviction of the speaker or listener can therefore only be a matter of "illusions," as E. W. Scripture once forcefully expressed himself.[1]

No mathematical point can be found on these curves of which it could be said: here this sound ceases and a new sound begins. This fact is not based on phonetic experience, nor is it a *result* of a mathematical analysis of these curves, but lies in the very concept of those curves, which have been obtained by the registration of natural processes. Such a process—of the duration of a second, perhaps—can naturally be represented graphically as a curve of, for example, 1 mm or less, or 1 km or longer, at will; therefore, it would be foolish even to seek a mathematical point at which one sound ends and another begins.

In addition, there is also a "phonetic" argument: for the production of a speech sound the cooperation of a number of different organs is necessary; or, more precisely, the teamwork of striated skeletal muscle-fibres which are arranged in a number of anatomically and physiologically differentiated muscles. It is highly improbable and, as Menzerath has shown,[2] in fact does not happen that all these muscles and organs move simultaneously from the formation of one sound to the formation of the next. However, if one group of muscles begins sooner than another, it not only cannot be stated *where* on a physiological or physical curve one sound ends and the next begins, but not even *when* this happens.

"Objectively," i.e., from the physiological and acoustic aspects, one can therefore only say that there are indeed portions of curves that correspond to perceptible speech processes; but these curves are continuums and hence exhibit no boundaries that allow a succession of discrete sounds to be separated and distinguished from each other.

By extracting the transition phases situated between two relatively stationary phases of the acoustic curve and by making the sum of these segments audible we can show that our auditory impressions of speech relies on the organization of our ear or on the structure of our auditory experience. With this experiment one would provoke auditory perceptions which escape use in listening to spoken language. This fact also supports the view that no science can be based on experiences founded on the "inadequacy" of our organs and experiences of perception.

With these last two arguments, it is a question of empirical facts: of facts known to us from the anatomy and physiology of the speech organs, from the structure of the sound-pressure curve, and from the character of auditory experiences. However, the linguist who finds himself confronted with a diversity of linguistic structures presupposes all this without having to demonstrate and differentiate it in each language. The concept of categoricality indeed does not correspond to the concept of Kantian categories, but is already established in his cognitive theory inasmuch as Kant distinguishes between a *quaestio juris* and a *quaestio facti*. As a psychological reality, the latter belongs entirely to the conditions for physical as well as mathematical cognition: the mathematical judgment does not receive its legal argument and claim to validity from the fact that it is thought, although they both presuppose that it must be capable of being thought. To this extent the sciences, without losing the diversity of their aims and methods, form a tissue of reciprocal conditions.

Some twenty years before the development of experimental phonetics—that is, before the appearance of Rousselot's *Principes de phonétique expérimentale*[3]—Hermann Paul pointed out in his *Prinzipien der Sprachgeschichte*[4] the linguistic situation just described: "A real analysis of the word into its elements is not merely very difficult but is actually impossible. A word is not a compilation of a definite number of independent sounds, of which each can be expressed by an alphabetical sign; it is essentially always a *continuous series of infinitely numerous sounds*, and letters do no more than bring out certain characteristic points of this series in an imperfect way. The remainder, which remains undenoted, no doubt necessarily reveals itself from the definition of these points, but reveals itself only up to a certain degree. This continuity can be recognized most clearly in the so-called diphthongs, which exhibit such a series of infinitely many elements."[5]

For the reasons given, the number of sounds represented by the curves in question cannot be ascertained with certainty even through a Fourier analysis. For example, the curve in Figure 4 contains six oscillations of a spoken *a* from the word "das"[6] in the question resulting from the speech-situation: "Was ist denn das?"

The first oscillations reveal three peaks, the last show four. We can see very clearly the fourth peak, which increases "gradually," or more precisely by steps, from oscillation to oscillation. The cause of this increase lies in the frequency position of the oral cavity formed for the *a*-position, which prevails also if the frequency of the fundamental is lowered. This illustration already suggests that one cannot decide from such a feature of a curve whether we are concerned here with a transition from one vowel to another—this could not be decided even on the basis of a harmonic Fourier analysis, of an automatic sound-analysis, or of a Fourier integral. This is true quite apart from the fact that the

Figure 4

different methods of harmonic sound-analysis and also the Fourier integral may be applied only on condition that stationary relationships exist for the investigated characteristics of an oscillation. It also cannot be decided where one type of curve begins and the other ceases. Two contiguous sounds thus cannot be physically segmented with certainty—in the sense that there will always be intervals of time, or portions of curves corresponding to them, of which it will remain doubtful whether they belong to one or the other sound.

The same objection to the possibility of distinguishing a succession of spoken sounds can be advanced with physiological arguments. For without doubt objections can be raised as to the concept of articulatory *position* if it is to guarantee the distinction and isolation of the sounds. Jespersen[7] particularly has shown—and this can be shown still more exactly in the context of X-ray movie-film investigations on the articulation of the vocal tract—that the concept of tongue position is to retain its importance for certain questions of the physiology of speech sounds; nevertheless the function of this concept must be restricted, and the articulatory position will finally have to be viewed as an exception to the movement of the articulating organs. The question of the articulatory positions resembles that of the formants of the different vowels. If we investigate, for instance, the tongue-position in isolated sung vowels, it is certainly not difficult to find for each sound a particular position and its particular formant-content. However, if we examine the movement of the articulating organs during a word or sentence on X-ray film or X-ray sound-film,[8] we see that from curves of motion thus obtained no definite judgment as to the type and number of the spoken vowels can be made, and it certainly can not be stated: this is where this sound ends and this is where that one begins.

It could be objected that it is not absolutely necessary to find this in effect undiscoverable point in the structure of these curves. Since even a glance at the curves allows us to distinguish at least on some occasions between phases of faster and of slower change, the former could well be understood as the portions of curves which correspond to what have been designated "transition-sounds," and the sections of slower changes would correspond to the real "sounds as such." This is supported by the similarity of these sections of the curves (for instance, of slowly spoken vowels) to curves that correspond to sung vowels. We could distinguish nucleic sections of those curves from transition phases and segment them to some extent. In this way we would arrive at natural foundations of discrete diversities of spoken sounds.

But even if we wished to support these arguments and claimed that with mathematical and physical procedures we *could* succeed in segmenting sounds which could be distinguished from each other, we would still have to reckon with the possibility that the ear of the listener arrived at a different number of different sounds than did the analysis of the curves. Indeed, it is possible that representatives of different language communities using a transcription procedure would arrive at different results. That, too, speaks against the usefulness of this subjective physical procedure for scientific ends and thus against the validity of experimental phonetics, which now as ever foregoes testing its concepts "experimentally."

The first objection runs as follows: the objective facts, on which the speech experience and the auditory experience are based, form continuums representable as curves; they do not allow of treatment as a succession of discrete elements, which would correspond to the articulatory and perceptual experience of discrete speech sounds. And more exact analyses reveal as the basis of these experiences characteristics of the physiological processes with which our perceptions are connected.

Paradigmatic Objection

At the paradigmatic level a new difficulty is added: if one considers, for example, the sound-pressure curve of a short narrative—which contains, let us say, a thousand individual units, which appear in the form of a thousand phonetic symbols—we would soon notice that we would find ever-new relationships from sound to sound or from one such nucleic section to the next. Even allowing that the distinction between phases of faster and of slower change could be carried out with certainty—which, as is well known, is not the case—not one single curve would exactly resemble another, even when restricted to its nucleic section.

By means of the artifice employed by us—that is, by renouncing definite boundary-points and restricting ourselves to border-phases—it would certainly be possible for us to distinguish and to count successive sections; so that, in the ideal case, we would reach a thousand sections of curves, which would correspond to the thousand phonetic symbols. Although it is possible, in a comprehensible text, to arrange those thousand successive symbols into a limited number of classes, a corresponding division of our thousand nucleic curve-sections would not be successful—and not so much because all these sections of curves deviate from each other (for one could, after all, say that of the written phonetic symbols also) but because we have no immanent criterion to determine which deviations are relevant for such a classification and which are irrelevant.

If we intended to discuss classes here at all, we would have to say: each such section of a curve has its own class, so that these portions are indeed countable, but not divisible into classes.

And what was just said of the physical sound-pressure curve is, naturally, just as valid in principle for the movements of the lips, tongue, etc., or for the curves by which these movements are represented.

With this objection, the psychological situation is, of course, rather different: if one hears even a few sentences of one's own or of a completely foreign language, one certainly has the impression that sounds and whole groups of sounds are repeated, and indeed repeated very often, usually. We cannot free ourselves of this impression of "identical" segments which not only can be counted, but also can be grouped into classes. However, on the other hand, the argument of experimental phonetics must be taken seriously, that these psychological identities do not correspond to any physical-objective facts, that we rather find in our example as many different curve-sections—and indeed so much more certainly, the more exactly we record—as physical units can be counted.

A further reflection is appropriate here: the number of classes that could be found in a recording of limited length would at the most be equal to the number of these segments it contains. But the reflection is not to be lightly dismissed that, with an increase of investigation material, a continuous increase in these classes would also have to be undertaken, particularly if we then went on to gather recordings of different languages and dialects. In this case, one must also consider that each particular speaker has his individual "speech-organ"; that is, a formation (peculiar to him) of the mouth and nose or of the timbre or of the distribution of the formants in his sound classes. But since between every two sound classes or timbre classes yet a third can be imagined, a compact continuum of sound classes would arise, so that frequency investigations would *only*[9] be possible with the help of artificially enforced principles of distribution. For the possibility of gradation in the movements of the speech organs, and naturally also in the sounds produced by them, is endless.

The second objection runs as follows: not only does each finite number of sounds have its own class, so that these sounds are not divisible into classes, but there are infinitely many such sound classes, so that we can speak only of a continuum of sound classes, which forbids not only a division into classes but also a sure distinction of classes.

Against the concept of set and its two criteria profitable for the investigation of speech sounds, two objections—one syntagmatic and one paradigmatic—should be raised, from the viewpoint of the natural sciences:

1. Against the distinction of speech sounds tells—from physical, physiological, and psychological viewpoints—the impossibility of fixing exact boundaries between each two successively spoken sounds; that is, the fact of a phonetic continuum of the spoken word.

2. Against the possibility of adding up sound realizations of a real conversation in a particular sound class, and against a natural establishment of sound classes tell two facts: First, every sound realization represents its own class, from an acoustic and physiological point of view. Secondly, there are theoretically—both physically and physiologically—an infinite number of sound classes; consequently there is a continuum of sounds; that is, a sound-class continuum.

2. SOUND CLASS AND SOUND REALIZATION

The above objections against the frequency investigations of linguistic segments and measurements of characteristics of spoken sounds or of suprasegmental structures are based on natural auditory, acoustic, and physiological facts and considerations, but neglect the difference between *language* and *speech*—that is, the fact that the *sound class* is neither a purely auditory or motorial quantity, nor an acoustic or physiological one, but a *linguistic* quantity. These objections thus overlook the autonomy and superiority of linguistics, which is alone qualified to elucidate linguistic facts by its own form of experience and by a procedure of proof peculiar to it; that is, to describe and to analyze them in a form which is suitable for the aim of linguistics, for the system of the languages.

Whoever is confronted by a written or spoken text and disregards the content of what is read or heard, in order to apply himself to the writing or the auditory impression as such, notices that such a text does not consist of ever-new letters or sounds, but that certain letters or sounds—and in longer texts finally all of them—appear repeatedly, with a certain frequency.

The fact that these symbols can be counted is not only of great practical importance for the whole of linguistics, especially for phonology in its widest compass—for phonemics and phonometrics—but it represents a problem in scientific theory, which pertains to the constitution of linguistics as a science.

It is indisputable that we have the knowledge of such frequencies of letters and sounds from experience. But this experience is different from that by which we distinguish two languages or language epochs or language strata from each other. The relationship between these two forms of experience can be easily shown: the idea that letters or sounds of the text of a language can be counted must be presupposed where languages are to be contrasted and compared. And therefore the proof procedure is different here and there: differences between two languages or dialects, for instance, are established with respect to experience—in the final resort, by reference to written or heard texts, ones that can be written or heard, and thus with respect to linguistic experience. In the other case, it must be demonstrated—disregarding the question as to how this knowledge has been obtained—that the concept of "accumulation" must be presupposed in order that such linguistic experience can be gained. In this sense, such an epistemological discussion is of an a priori type: it reveals in the linguistic judgment the conditions without which it would not be a linguistic judgment nor a proof procedure, different from other types of judgment or other sciences. Such a theory of the sciences thus makes a contribution to the question of the system of conditions which characterize linguistic experience as scientific experience and distinguish it from other scientific ways of gaining experience and of procedures of proof in the system of the sciences.

In this sense, an investigation of the problem of the accumulation of letters or sounds is an epistemological a priori "discussion" in the Kantian sense—an investigation which extends far beyond the problem of the accumulation of linguistic segments, into numerous other sciences.

If linguistics presupposes for its work a diversity of languages and dialects, language epochs, and language strata, it means by this structures valid in historical, geographical and social dimensions, which as such characterize a communication-community of people. Both as speakers and listeners, members of this community, when they communicate linguistically, are guided by what pertains to the language of such a community; that is, by what characterizes the language usage of this community in contrast to others. In this sense, the language usage is *normative*. "Speech proceeds according to the rules of the language." "Only through the medium of speech can we approach language." [1] This relationship is reciprocal to some extent: Since the language community in speaking and listening is guided by the structures valid in its geographical, social, and historical dimensions, linguistics must seek and study these rules in their usage, in their application by the language community; that is, in their *realization*. Everything else is linguistics of the letters of the alphabet and has only preparatory significance. This judgment is valid for a not inconsiderable part of linguistics to date, despite all its great successes and achievements.

Linguistics itself is indeed no longer normative, as the linguistics of the 18th century was. However, the objects whose validity and (historical) structures it investigates are for their part normative and consequently not directly perceptible structures; that is, they are neither visible nor audible. For what we can hear is only the *realization* of the language usage valid in a language community as it is guided by those rules and norms. These realized structures, however, have no validity in geographical, social or historical dimensions.

The sounds of a particular language are nothing but the speech norms of a language community. It is with reference to these that the perceptible structures of spoken language are described, distinguished from each other, and arranged. They themselves, however, are norms and as such not perceptible; but on the other hand they cannot be exhaustively defined scientifically when detached from their realization. This insight gives the invention of record, sound-film, X-ray sound-film, and tape its scientific weight—which was, however, until recently not sufficiently recognized either by linguistics or by experimental phonetics.

It is not that there are, first, perceptible sounds and, secondly, imperceptible sound-norms. *The same* linguistic structures can be considered from the viewpoint of their non-evident, normative structure— that is, from the viewpoint of linguistics—and also from the viewpoint of their realization in a particular, unrepeatable conversation. These concrete sound-structures, again, can be considered as objects of linguistic psychology,[2] as objects of linguistic physiology (with reference to the sound-producing organism),[3] and as objects of linguistic physics.[4] But a linguistic arrangement of these concrete structures is only possible with reference to the structure of the phonemes of a particular language.[5] This structure is handed down in a particular geographical and social dimension. The first linguistic question therefore is: Which of the characteristics of those structures are handed down? In this way diachronic considerations are brought into purely synchronic tasks. This guarantees the link between these two viewpoints.[6]

In the discussion of the linguistic aspects of language, we deliberately disregard the content of what is spoken—hence one must certainly avoid bringing the choice between form and content into the matter. We do so first because, under the pressure of these alternatives, the objects of linguistics as well as those of literary criticism[7] and the history of literature[8] would belong to the "form." In both cases, problems of meaning also play a role—for example, the questions of semantic fields,[9] which differ from language to language—their social and geographical validity and their variation and issues, of aesthetic experiences of meaning. These are problem areas, which avoid this choice. And secondly we shun the choice between form and content because the field of "content" coincides with the objects of historiography, and the system of contents is the system of those objects. Although the aspects of linguistics, literary criticism, and historiography belong closely together and cross-fertilize each other, the aims, concepts, and procedures of proof of these three disciplines must be kept strictly apart; the aspects from which at times a "source" is approached and examined must be precisely distinguished. The system of languages, the system of (linguistic) works of art, and the system of documentary history are three different research aims with three different types of validity and tradition. Even when they are held together by the concepts language and source, this cohesion—like that of mathematics, physics and chemistry—is a connection of the forms of scientific procedure, hence of cognitive theory, and not a connection of objects of *one* science, not a "content-based" cohesion. That would be a new form of science, as it has been postulated since the 18th and 19th centuries, now for the natural sciences, now for the humanities—the latter particularly in Germany since Romanticism.

If it is, for linguistics, a matter of *what* the speaker has said, then the response to this question is in the last resort a matter of the language community within which that speaker is understood and in whose name the linguist answers this question. It is not yet decided and requires careful investigation, whether the question as to the content coincides with the question as to which characteristics can be perceived and deliberately produced. In the learning and speaking of a particular language, certain regulating mechanisms are used; they are definitely related to perception and deliberate production, but need not be identical with them. It is well known that the perception and deliberate speaking of foreign languages need not coincide; this can be demonstrated experimentally at any time. It is, however, quite possible that the same phenomenon is also true for one's own language: it is even probable that we can perceive and deliberately imitate only what we have learned to pay attention to in the course of learning, practicing and using it, and that the characteristics which we have actually learned to disregard during this process at first evade our perception and deliberate speech. This is true for both the characteristics of the spoken sounds and differences between these characteristics.

And so a particular tendency of research manifestly approaches language with the task of seeking and determining differences between linguistic phenomena, which from a *linguistic* point of view are declared identical. It is evident that this formulation, which includes the conditions of several sciences—psychology, comparative biology, and physics—amounts to *statistical* tasks: they concern the distribution and load of linguistic classes; these methods of counting and measuring thereby become linguistic procedures. The traditional, geographically and socially defined sound-norms—which are the subject of linguistics—are here understood as *sound classes* that, although they are not themselves perceptible, represent a regulation of the perceived and of the spoken *realizations*.

In this sense we will always distinguish in the context of the following investigations between (concrete) *sound realizations* and (traditionally transmitted) *sound classes* which—as is to be shown—do not coincide with the *phonemes* of phonology. In German, for example, the *ich-* and *ach-*sounds represent the same phoneme, but—as so-called *allophonic variants*—they represent two sound-classes, which as such are again not identical with *perception experiences*, but are the norms of these.

The fact that a particular "sound" appears in a particular language with a particular frequency, can mean two things: either that the sound class in question appears frequently in individual normative sound-combinations (words) of the language in question (as in a phonetic dictionary) or that (few) particular words are realized frequently in texts or conversations, so that this sound class becomes particularly loaded through these repetitive realizations. The partial dependence of the second question on the first is obvious. However, the two questions have to be distinguished. In keeping with what has just been explained, the totality of the sound classes—that is, the phonetic inventory of a language—will be called the *sound-class stock* of this language; the totality of successive sounds of a concrete conversation or speech will be called the *stock*, the *quantity*, or the *totality of realizations* of the sound classes of this audible speech or of that audible conversation.

3. THE SOUND UNIT (SOUND SEGMENT)

Linguistic Unit of Sound

The sound class is not a fact that arises and disappears in the act of speaking, but a norm that is conditioned by the structure of the language concerned, and as such is transmitted from individual to individual, from social stratum to social stratum, and from generation to generation, in a geographical area. In the course of its history, this norm undergoes specific transformations in the context of these structures, which may have been transformed; speaker and listener subject themselves to that norm if they wish to communicate linguistically. Auditory, physiological, and acoustic facts are three related but distinct aspects of the realization of such normative sound classes—but are not the classes themselves.

In this structuralistic-synchronic dimension, we must again ask and test the question as to the possibility of distinguishing and segmenting the speech-sounds; that is, the question as to the sound class. In any case, this question is from the outset restricted in several ways by the structuralistic point of departure. It is not a structurally relevant problem to consider all perceptible and recordable or calculable sound-nuances—for only the *norms* of these auditorily, acoustically, and physiologically comprehensible realizations are of linguistic significance in the stricter sense. Nor is it the task of linguistics to treat the sound classes of all or even of several languages without reference to their structural organization. The linguistic question can only be phrased: Is there a finite number of distinct sound classes in each geographical-social-historical area of a language?

Before the answer is given, we must determine the scientific dimension to which this question and its answer belong. Can linguists decide the question by referring, for instance, to the finite number of distinct sound-classes in this or that language? Can the question really be answered at all by referring to actual conditions determined by experience? Could we not object to such an answer on the grounds that, though linguistics operates with a limited number of sound classes in this or that language and perhaps in all known and investigated languages, the question nevertheless remains unanswered whether and by what right linguistics is qualified to do so? Therefore, by asking *expressis verbis* about the procedure of linguistics, it becomes clear that this is not a linguistic question, but a methodological or epistemological one, which of course cannot be countered by linguistic arguments, but only by arguments taken from cognitive theory.

If this is recognized, then the question can be formulated anew and more precisely: Can it be shown that it is part of the concept and conditions of linguistics to work with a finite number of sound classes in the analysis of the structure of every language in every stratum and every epoch? Is linguistics compelled to do this by virtue of its own method, in order to be able to determine and to master its subject at all?

This problem can only be broached here by means of hints. It could be answered: Linguistics assumes that the structures which it investigates are norms of communication; that is, norms by which speaker and listener are guided, when they wish to communicate by means of language. These norms must therefore be capable of realization in speech (that is, of becoming perceptible natural forms); these in turn can again be understood—referred to the same norms. If there were infinitely many such norms, which varied from one another by a quantity no longer distinguishable by the ear, then communication would be suspended; these norms would no longer be able to perform their task of serving communication, and these forms would no longer be subjects of linguistics—they would no longer represent language. It could equally well be demonstrated that the concept of sound change is based on a finite series of normative sound classes in each phase of linguistic history.

Only with reference to such a finite number of distinguishable sound classes can we speak at all of the infinite variations of realized speech sounds in auditory, acoustic, and physiological respects.

And since linguistic sound classes are distinguishable types of speech sounds, the application of a finite number of written symbols is justified. Certainly no reference is thereby intended to the adequacy of any particular traditional or modified orthography, nor to the utility of a particular phonetic transcription system. What is meant, however, is that it is nevertheless a justifiable scientific task to look for appropriate transcription systems using a finite number of phonetic symbols for each language. The justification of this "abridged" method of recording is not to be denied, but we should ask whether all the structurally relevant differences between sound classes are represented in this particular phonetic transcription system—whether, for instance, certain transition sounds are significant in communication so that special symbols should be sought for them. Such reflections do not deny the justification of finite, discrete symbols but presuppose it. Alphabetical script is not something added externally to language, it is implied in the concept of language and it is a suitable form for representing it: "Thus writing is connected basically with language.... The tendency towards writing is always posited with the fact of language, writing and language are in this basically related reciprocally, but always in such a way that language remains basically the primary thing and not writing. In the tendency towards writing, language accommodates itself to the continuity of the community. However, community follows from the motive of communication; or, in other words, this motive includes a community. And communication, we know, means 'objectivity.' So this is the legitimate root of the term 'language.' " [1]

Since the distinguishability of symbols is usually one of the conditions of communication through symbols, and not merely linguistic communication, we find the same situation also in all other areas of symbolic communication. For example, if we wish to create an unambiguous communication by means of a series of colors, then these colors must be unambiguously distinguishable to the whole group of people communicating by means of them. It is irrelevant here whether the red used in a concrete case is physically completely identical with that prescribed for communication. It is sufficient for it to be recognized as the prescribed red; that is, it must be classified with the color-class in question and distinguished from other colors. If the colors employed had to be produced anew in each case, then it would appear, for example, that communication is possible and indeed is carried out with various reds, which can all be precisely defined physically. However, these shades of red would presumably be grouped around a number of maximum frequency if we were to treat their physical values statistically. For every community there could always be only a limited number of such maximums, if communication is not to suffer or to be suspended. That naturally does not exclude the possibility of these maximum frequencies lying at quite different points of the spectrum in different communities. Therefore, an infinite number of such maximal points—corresponding to the sound-classes—could be imagined for as many communication-groups as desired.

To say that linguistics must presuppose a finite number of distinguishable sound classes for every language in every phase of linguistic history, is, of course, not to say that these sound classes would have to be capable of forming a mathematically unequivocal order—similar, for instance, to a number series. On the contrary, it was said of these sound classes that they were not any particular things—detached from the concrete realizations of perceptible sounds—but merely the expression of a particular way of considering concrete speech. These concrete structures are themselves to be considered and investigated from the auditory, physiological, and acoustic points of view. None of these forms of arrangement of scientific investigation can be ignored in the ordering of the sounds of sound classes.

In experimental phonetics and the traditional linguistic classification or ordering of sound classes, the physiological point of view has undoubtedly been in the foreground. The external reason for this is that previously neither acoustic[2] nor "auditory" analysis provided linguistics with such essential organizational factors as the physiological investigation of sound production. But even the physiological arrangement of the sound classes—into velar, palatal, and labial stops, etc.—has in itself no serial character in the sense of mathematical order and can in itself have none, since the specific arrangements of the organs represents a "presential centralization,"[3] which excludes the concept of series. This means, however, that there are and must always be several possibilities of arranging the sounds from a physiological viewpoint; for example, they can be arranged according to the place of the essential articulation (even this concept is by no means as unequivocal in all cases as it is often taken to be), according to the activity of the vocal cords, the position of the velum, the rounding of the lips, etc. These possibilities of manifold suitable arrangements for the sound classes are not a deficiency, but follow from the physiological and anatomical structure of sound-formation. Certainly we may say that in all these "phonetic" attempts at arrangement, the viewpoint of structural linguistics has often receded too far into the background, in comparison with the physiological viewpoint.

When the finite number of distinguishable sound classes is established for a particular language or linguistic epoch, then it must still be investigated whether these sound classes can be distinguished and counted in the speech chain. We must ask especially whether the concept of the *transition sound*, as defined by Sievers, does not destroy the possibility of counting the sounds in the spoken word.

But here too we must call to mind again the difference between the linguistic viewpoint on the one hand and the auditory, acoustic, and physiological viewpoints on the other. The latter concern only the individual realizations of the sounds in the context of the unrepeatable conversation. They have *as such* no linguistic validity but are at best perceivable events. They are *more* than such events only because speaker and listener associate them with the finite number of sound classes. Only these sound classes are immediate objects of linguistics. It is they to which the speaker refers, when he expresses himself verbally. The question as to the possibility of counting the sound classes in the spoken word should consequently not read: Are any transitional sounds audible, or are any transition phenomena physiologically or acoustically present? But it should read: Are the sound intentions of the speaker calculable in the context of the sound classes of his language? An example may illustrate this question: If in High German (such a stipulation is necessary, if the investigation is not to cease being a linguistic one) we pronounce the word "das," we have—inasmuch as one substitutes only one element at a time—only three possibilities of deliberate alteration with respect to the sound-classes. We can say "was" or "des" or "dann," and nothing more or less; that is, we can deliberately alter three sound classes.[4] Everything that we do beyond this is no longer High German and so, in the context of a conversation carried on in High German, out of place and therefore irrelevant from the viewpoint of language—and linguistics. What can be heard, or what physiologically or acoustically occurs between the /d/ and the /a/ or the /a/ and the /s/ does not change the fact that this is a word with three commutable elements. Naturally this is not to say that in the pronounciation of a word with three sounds we are concerned with three articulatory acts of the will. This question would belong to the psychological dimension of speech and not to the specifically linguistic dimension of language. Rather we must presuppose psycho-physical control mechanisms, which can scarcely be analyzed without the consideration of linguistic problems.

The objections to the distinction and segmentation of speech sounds can be opposed by a series of propositions, on which all investigations of the frequency of speech elements and all acoustic and physiological measurements of speech acts are based:

1. Every language has in every epoch a finite stock of sound classes, or the sound-class collective. There is no continuity of phonetic sounds for an individual language, but the stock of sound classes of every language represents a discrete quantity.

2. Considering the fact that the linguistic norms are carriers of meaning,[5] the sound-class stocks of different linguistic epochs of the same language are connected with the phenomenon of the change of sound classes.

3. The system of sound classes is conditioned primarily by linguistic factors and secondarily by auditory, acoustic, and physiological ones, and consequently it reveals no serial character.
4. Every conversation has a finite stock of distinguishable sounds: the sound stock, the number of sounds, or the sound collective. From the linguistic viewpoint, there is no continuity in successively spoken sounds. Therefore, all the sounds of a conversation also represent a discrete quantity.
5. The principle of arrangement of this sound collective is the system of the sound classes of the language in question. By this the possibility of accumulating sounds in sound classes is guaranteed.

Auditory Sound-Unit

If it is established that the two criteria of the concept of set—distinguishability and segmentability—can be applied to speech sounds, then we must still examine how these considerations work out in detail in auditory, acoustic, and physiological respects. From earlier discussions it appears that there is only *one* procedure which here aids progress: linguistically relevant sounds are present in a conversation or a tape-recording of it. What part of them is to be symbolized in script, and how are these sounds expressed as distinct countable and measurable units at the auditory, acoustic, or physiological levels?

There are two psychological definitions of speech sounds; namely, one based on the auditory experience of spoken words and the other founded on the speech mechanism; that is, on the cognitive experience associated with speaking.

As experience is a fact and at the same time the principle underlying the object which has been experienced,[6] the deliberations on these two questions also have to be treated in accordance with the real and the ideal structures of the human mind. "To experience something" includes the tendency towards verbal formulation because knowledge and thought[7] can be appropriately expressed in language. For an analysis of the speech experience, it is necessary to focus our attention on the possibility of passing from thinking to the verbal formulation of the thoughts. This also means that we can experience a temporal sequence of sounds, whereby not only the first sound has an effect on the second, but also the second on the first; in this way syllable, word, clause, or sentence become units of different orders. The experience of such perceptions can be analyzed only after they have been caused by a suitable arrangement of the psychological experiment. If this aspect is excluded, only the intellectual or cognitive relations between thinking and speaking can be investigated. The role played by the actual experience of the succession of speech sounds and by their actual articulation is well known, but it certainly requires more penetrating investigations within the context of articulatory measurements.[8]

A temporal succession of speech sounds consists of countable items only from a linguistic (phonemic) point of view. This however does not mean that the psychological experience of these speech sounds must consist of countable items.

> Only objects can be counted. Experiences can be counted only inasmuch as they can be evaluated as objects in a certain sense; and this is correct. Its conditions appear in the important connection of every experience with the function of "consciousness" or "knowledge." [9] . . . To be a psychological object means to belong to a temporal order, or to be able to belong to it; in this order every member together with a certain number of preceding or following members forms *one* unit of experience. That something can be understood means it forms a unit of experience together with the other members and that it is included with these in *one* "intentional" act.[10]
>
> A variety, with reference to the unit of experience, means precisely a known or one that could be known and is formed in an order of experienced time. And vice versa: To be formed in the order of an experienced time means to belong to the unit of a connected experience.[11] . . . The psychological conditions for calculating also constitute the possibility of counting the "object," and they do so precisely by their dependence on the structural law of meaning.[12]

Hence, by an analysis of thinking as it occurs in speech or articulation, we pass to an analysis of language perception—both of which must possess basically the same structure: the structure of the psychological process.

It remains only to add the following deliberation in respect of our problem: Wherever an experience of a variety is under discussion—and that is the case at both the articulatory and the auditory levels—certain demands have to be made on the object experienced. This object must itself be suitably organized and the elements in its variety must be distinguishable. The better they can be distinguished, the easier will be the possibility of perceiving this organization. No doubt there are borderline cases in which these experiences will not occur at all—think, for example, of a cloudless blue sky, or of an extended musical note. In such cases either no object-oriented experiences occur at all—the note is "overheard"—or elements of these objects appear in devious ways, for example, by experiencing them in our own bodies.[13]

These deliberations contain references first to the importance of an easy distinction of the sound classes and secondly to that of actual realization. Here too it can come to borderline cases where the distinction of sounds is no longer felt—for example, in children's speech and in various pathological phenomena—and where for this reason nothing is understood any more. In this case the speaker loses his ability to think in language as well as the object reference of his thinking altogether.[14]

The recognition of distinctive speech sounds necessitates the knowledge of the stock of sound classes, and it also requires that the deviation from the norm should not exceed a certain proportion in the individual case.

If we can experience the phonetic articulation of a word or a sentence structure, then these phonetic experiences can be counted within the limits indicated above. From the viewpoint of counting them, these perception experiences shall here be called "units of phonetic perception," and the acts on which the experience of articulation of the word is based shall be designated as "units of phonetic intention." We have pointed out the basically similar structure of these two psychological "units." We discussed earlier the fact that an organization of such units can only be based upon a finite number of different sound classes when we think in terms of articulated thinking and of language perception; that is, in terms of communication. The sound classes are related to these psychological units just as the tasks of linguistics and of psychology are related to each other; they are based on different assumptions, though certainly closely connected and conditioning each other reciprocally. And similarly the large variation of psychological sound units (of one or several languages) and the finite number of linguistic word classes within the same linguistic epoch of a language reciprocally condition each other.

However, the psychological analysis of perception experiences will under certain circumstances encounter structures which cannot immediately be associated with a certain sound class, but the context may suggest the conclusion that two sound classes must have been intended in this perception, which cannot be further differentiated. Such experiences, which can no longer be directly projected onto linguistic (phonemic) facts, may be called "phases" as distinct from the "units": they are "phases of phonetic (auditory) perception" or "phases of phonetic (articulatory) intention." It will occasionally happen that the linguistic units do not precisely coincide with the psychological phases. This will always be the case when the phonetic realization has deviated from the linguistic norm to a certain degree. The "phase" therefore represents a principle of segmentation as used in a more-or-less independent psychological analysis of linguistic experiences and the "unit" stands for the completed projection of psychological and linguistic orders onto each other.

The articulatory or auditory distinction of two phonetic realizations, however, has its limits. Perhaps it varies from person to person, but certainly these limits can be extended by observation and drill; for example, a glance at the oscillogram or (better) at the sonagram reveals differences even between the two sounds /aː/ articulated if the word "father" is spoken twice. If the word is repeated by another person, we can certainly also notice the difference in timbres; sometimes this is the case only if we are attentive to this, sometimes the difference obtrudes upon the similarity (especially on the telephone, when we do not see our partner and perhaps have even to guess who it is). If, however, the word is repeated by the same person "in the same manner"—that is, "equally loud," "equally happy," "equally fast," etc.—then no further auditory differentiation is possible. The words that are the same on the linguistic level will also produce the same auditory *impression*, even if their sonagrams or X-ray films are different. The same is true *mutatis mutandis* of the articulatory limits of precision in the production and reproduction of the "same" word. Here then linguistic sound classes (which coincide with phonemes, but can also be variants of "one and the same" phoneme) and auditory or articulatory sound units coincide.[51]

It is doubtless important, for the limits of perception and the limits of sound drills, that these auditory units of the twice-spoken word have the same function in the word and consequently are the same linguistic sound class. And the identity of their linguistic function is also of decisive importance for our investigation of two sounds in two different words, and by no means in the same context; for example, in the words "beat" and "seat", but also in "meet" and "mean" or in "seat" and "lean." In these instances we can hear that the phonetic environment is different, but the difference of the phonation of these /iː/ themselves is no longer perceptible. In this case phoneme, sound class, and auditory perception unit of the first two words coincide; while their differences may still be established by sonagrams, or at least one can always investigate whether there are differences between the sonagrams or oscillograms of these words. It is obvious that such problems can only be solved statistically as is required in each separate instance. The limits of psychological differentiation are precisely determined by the fact that the same speaker *intends* to pronounce the /i:/ in the same way not only in "beat" and "seat" but also in "beat", "lean," and "wheel," and that the listener *must* also understand this if he directs his attention to the *percepible*, but usually not perceived, phonation.

The following fact supports this: Singing is distinguished from spoken language first by differences of intention and secondly by the differences in subject matter. The singer—and the person hearing the singing—attends to the phonation in a different way from the speaker and the person listening to speech. The singer *wishes* to sing in a particular way, while the speaker primarily *intends* this or that and therefore (often without noticing it) speaks in one particular way or another. So, if the singer wishes to sing the second of two successive syllables on a higher tone than the first, he "raises" the note to the higher position; that is, he raises it beyond this position, and then he lowers it to the desired position, as pitch measurements of sung sentences have shown. This variation can extend to as much as half the duration of the second note, and longer. Indeed, strictly speaking, the singer's tone never ceases to "vary," because only in this way can the singer maintain the note at the same pitch, since he checks the pitch by listening. The rise and the variation is only heard if it exceeds a certain amount, which presumably is determined less by its objective quantity than by the singer's intention; if the person singing *wishes* to raise it—as is sometimes practiced in sentimental singing with vibrato—it is also heard. It is very probable that we could hear, or learn to hear, this pitch variation if, like the singer, we had learnt to pay attention to this. At the level of phonetics, the *social* character of language appears particularly clearly. Since the person singing does not intend the pitch variation but the interval, the listener also perceives just this intended interval, not however the variation that is actually sung.

The same applies to the contact between speaker and hearer. One could say, the listener would hear—within the limits established by custom and practice—not what he "actually" hears (what impinges on his ear—and he could perhaps learn to hear this consciously) but what the speaker intends; and the speaker "says" what he intends to say and what he wants the listener to hear.

The possibility of learning to speak and of communicating verbally depends, without doubt, on these control mechanisms; this is true for the subconscious acquisition of speech by very young children as well as its conscious acquisition in later years. These issues are part of the problem of the distinction of linguistic classes and of psychological units. Indeed, they justify the association of the sound unit with the sound class—a concept which represents the central concept of quantitative phonetics. Association is the bracket which allows viewpoints that must be distinguished in the investigation of speech to be united again, and indeed with the use of statistical methods.

How little this requirement has to do with biologisms or the statistical methods used in biology and is, therefore, an approach on a lower level when compared to the linguistic task, may be illustrated by the following quotation from Viktor Kraft's[16] *Erkenntnislehre*:

> The purpose of language—namely, objectivation and communication—demands that a sign be repeatable at will and common to a majority of persons. But in reality a sign consists of an individual acoustic or optical form, which is present at a particular point in space at a particular point in time. The real signs are different from one another, not only with respect to space and time, but also in their form. The word "the" appears many times on a page, and in different books in different letter types, and in handwriting its forms differ still more, even with the same person; in the writing of different persons its forms can deviate to an astonishing degree. Similarly this holds true in speech. And these forms all represent the same word. On the other hand it is one word (the article "the") which appears in all these different forms.* This one word identifies the actual variety of its individual forms. However, only the many individual signs, the spoken and the written words, are real; the *one* word is a conceptual unit. It is based not on a real identity but on the fact that the individual signs resemble each other, that they constitute a *class* of forms. Only certain characteristics or their relations**—in short, a type—are important. Such a type is a class of forms which is determined by their resemblance to an ideal form. A particular sign represents a linguistic sign only by virtue of its membership in a certain class of forms and only as such can it justify the two demands made on it.

* Already expressed by Peirce in *Prolegomena to an Apology for Pragmaticism* (Monist, 1906); discussed in more detail by Pentillä and Saarnio: "Einige grundlegende Tatsachen der Worttheorie . . ." Erkenntnis, vol. 4, 1934), p. 38f. See also Carnap: *Introduction to Semantics*, 1946, p. 5; Reichenbach: *Elements of Symbolic Logic*, 1947, p. 4; Pap: *Elements of Analytic Philosophy*, p. 311.

** Described by K. Bühler in "Die Axiomatik der Sprachwissenschaft," 1933 (Kant-Studien, vol. 38) as "abstractive relevance." A very thorough and precise exposition of the sign as example and class (as also of "expression" and "statement") is in Dürr: *Lehrbuch der Logistik*, 1954, pp. 5–6, and in Saarnio: "Betrachtungen über die scholastische Lehre der Wörter als Zeichen," 1959 (*Acta Academiae Paedagogicae Jyväskyläensis*, XVII).

The sound classes contain not identical but identified auditory quantities and, therefore, also identified acoustic and physiological quantities. Listening and speaking are not passive functions; they are productive processes, and the word which we hear is a unit of meaning which we produce ourselves. Only when quantitative phonetics takes these ideas into account does it perform its linguistic task and cease to be a "natural science." Nineteenth and early 20th century phonetics was mistaken in its desire to be a natural science.

We can sum up as follows:

As it is a linguistically established fact that we can distinguish and segment, and also calculate and measure speech-sounds, both the auditory and the articulatory experiences can be organized in calculable units; that is, in "units of phonetic (auditory) perception" or in "units of phonetic (articulatory) intention." If a coordination between the linguistic norms and the experience of hearing or speaking is not successful in the individual case, then this experience is divided into equivocally calculable "phases of phonetic perception," without altering the fact, however, that coordination and calculation are basically possible.

Physiological and Acoustic Sound-Unit

Accordingly we must now probe the relation between the calculable sound classes and the movements of the organs of speech—whether they are measured directly by means of levers or by X-rays or indirectly by recording the reaction of the air that is emitted by mouth and nose. In other words, the linguistic division of the spoken words into segments, and particularly into sounds, is to be sought in the completion of these movements. For this inquiry it is necessary to consider carefully the task and the limits of recording.

In linguistics, particular places of articulation of the various organs of the vocal tract—especially of tongue, soft palate, and lips—are distinguished; these places of articulation are associated with particular speech sounds.

With the usual pronunciation of [pa] or of a similar sound sequence, in which [p] is initial, the closure of the lips is, from a linguistic point of view, irrelevant; they may have been closed much earlier. It seems that the opening of the lips is the important movement. On the other hand, for the pronunciation of [p] in final position, in [ap] or [ip], the lips are almost always opened again after a brief closure, but this release is by no means necessary; in this case the decisive action seems to be the closing movement. If [p] stands medially, as in [apa, ipi] etc., both movements occur of necessity; but if we take combinations such as [æmp] (as in "lamp") or [mb] (as in "ambassador"), then we easily realize that the lips close with the formation of [m]; the closure continues from the [m] up to the following [p] or [b] without any interruption. In other words, in these cases [p] and [b] are articulated in the same manner as in the cases in which they were in initial position and in which the speaker had closed his lips beforehand for one reason or another. If we revert the sound sequence to obtain [pm] or [bm], then the opening movement for [p] and [b] is eliminated; or, more precisely, it is postponed until the following [m] has been pronounced. And finally, if we imagine a sound sequence [ampma], which often occurs in the rapid pronunciation of a German word like "Amtmann," in which the /t/ in rapid speech is realized as [p], thus [ampman], then we will find that in order to produce the [p], we need neither close the lips (since they are already closed), nor open them (since the following sound also requires bilabial closure). However, it would be foolish to distinguish all these cases, and to assert that we had a type of [p] in which only opening was relevant, another with only closing, a third with both movements and a fourth with neither. It will, therefore, be most natural and adequate to seek the really characteristic element of [p], or its distinctive feature, in the element common to all these cases; namely, that the lips completely interrupt the air stream at a certain moment. The essential feature therefore is not the movement of the lips, which depends on the environment, but the position, the "closure" itself. The same applies to [m]; what is common to the various /m/—the [m] in [ma, am, ama, amba, abma, abmba]—is a certain position of the speech organs in which the bilabial closure is contained as an element.[17]

A general exposition of the relations of such positions, and of the vowel systems built up on them, is given in the new edition (1923) of Viëtor's *Elemente der Phonetik des Deutschen, Englischen und Französischen*,[18] executed by E. A. Meyer, wherein it states:

All the vowel-systems discussed above suffer from the disadvantage that they are largely based on inexact observation and subjective evaluation, particularly with regard to the forms of articulation. A more precise determination of the vowel systems of educated German, English, and French is to be expected from experimental phonetics, which aims not only at the investigation of acoustics, but also of the articulation of the sounds.[19]

E. A. Meyer points out in particular the importance of investigations with the help of the plastographic method, which allows the articulation of the tongue against the palate to be represented. As he reports in his *Untersuchungen über Lautbildung*,[20] the curves obtained by this method are fully adequate to give us information on the articulation of vowels of the front series as also of the so-called mixed vowels, and finally of a whole series of consonants.

In 1905 E. A. Meyer took X-ray pictures of the articulatory position, with intensification of the outlines of the tongue and palate by means of suitably attached lead plates; he investigated by the plastographic method the relevant sounds in German (Northern, Central, and Southern German), Dutch, English, Swedish, Norwegian, French, and Italian. By this method, by means of which a series of the current views on the positions of the tongue were disproved, he reached essentially the same conclusions as by the X-ray method. Against objections he asserts "that neither can the tongue—even diagrammatically—be regarded as a point-like structure, nor does the space in which this point-tongue would have to move have in vertical section the form of a right-angled coordinate-system." "Both the X-ray method and Meyer's plastographic method" mean "a decided advance, in that instead of an (ideal) point the whole center-line of the tongue's surface is defined, and by direct mechanical means. It is clear that the configuration of the articulating tongue is thereby not yet perfectly produced. For this would require the definition of the behavior at least of the whole surface of the tongue." [21]

The X-ray or the plastographic methods ascertain the position of the organs for a particular sound which is spoken or sung for the purpose and the duration of the investigation. It has been pointed out in various connections, especially by Scripture, that there are no "sounds" and "sound-positions" at all; that the old concepts of the speech sound, syllable, metre, were "illusions," which possess "no real existence"; and that phonetics must attempt "to build up new concepts of reality on the experimental results." [22]

Essentially, his two arguments against the "reality" of the concepts of the sound and the articulatory position are the X-ray film of Gottheiner and Gutzmann, and the physical analysis of the sound curves. In his review of Gutzmann's publication[23] he writes:

> The impression made by a film of this sort is overwhelming. The organs of speech do not stand still for an instant; each speech act represents the sum of the movements made by all organs of the mouth, throat, larynx, etc., and this summation is enacted within time. Articulatory positions do not exist at all. One realizes immediately that the physiology of articulation as it has existed hitherto can only be a false doctrine and one awaits in suspense new developments.[24]

Without doubt, it is Scripture's opinion that, along with the "false doctrine" of the articulatory positions, the "illusion" of the speech sounds must sink into the earth, brought low, inter alia, by the X-ray film of Gottheiner and Gutzmann.

With regard to his thesis that phonetics must base new concepts of reality on results obtained experimentally, it must be objected that the investigations carried out by E. A. Meyer are also founded upon results obtained by means of phonetic experimentation. It was none other than Meyer who, by means of palatograms and X-ray photographs, contrasted experimental investigation with "imprecise observation and subjective evaluation," to say nothing of the fact that it was his experiments that enabled him to state the articulatory positions of speech sounds more exactly. The situation, if the X-ray film as such can be regarded as an experiment, would thus seem to be simply one experiment against another; and if Scripture wants to toss overboard "the physiology of articulation as it has existed hitherto," his opinion can only be that the experiments upon which he is based are relatively more conclusive and that those on which the physiology of articulation to date depends are erroneous, so that we are obliged to modify any views based on these latter results.

How, then, does he modify these views? He disputes the "reality" of speech sounds and articulatory positions; that is to say, he does not modify, but rather disputes the reality of the *results* of experimental phonetics. Since, however, he cannot dispute the reality of perceived sounds, palatograms, and X-ray photographs without also destroying the foundation of his own proof, it is to be assumed that he is not in fact out to dispute the reality of whatever sounds may be spoken or sung during the recording, nor that of their articulatory positions; but rather that he means something else altogether. What he is disputing, obviously, is the existence of similar sounds and articulatory positions during the process of speaking; that is, the justification which we have for applying results thus obtained to conditions differing substantially from the more or less artificial conditions of the experiments of earlier experimental phonetics. In the light of this, the crux of his whole argument can only be that results obtained with the aid of palatograms or X-ray photographs should not be granted general applicability without further consideration, that in fact new experimental methods will have to be sought as regards investigation of the speech process, and that the X-ray film will represent one of these new methods.

This simple solution to an apparent contradiction could only be regarded as satisfactory if it were to be assumed that, in point of fact, Scripture was only attacking either these experiments or their general application. That his criticism goes further than this can be clearly seen in the fact that he does not at any stage point to concrete deficiencies in the testing arrangements used in old-style experimental phonetics, arrangements such as the process suggested and used by E. A. Meyer, whereby small lead plates were fastened to the gums and tongue. He rather places his main emphasis on the experimental method as such, which he could not do if he merely wanted to recommend some other experimental technique. The fact that he speaks of articulatory position and speech sounds chiefly in a general way reveals that he is really objecting to certain concepts of linguistics itself.

We must now examine how far the concepts of linguistics are in harmony with those of E. A. Meyer, or to what extent Scripture's arguments may be regarded as conclusive and free of ambiguity—if, as seems to be the case, they are at odds with the former.

In this connection we need to remember that—to the extent that the method determines the object (that is, the requirements contained in the overall concept of a science constitute the object of that science)— the reality of scientific objects can be judged only within the framework of that science which has felt obliged to introduce these concepts during a certain stage of its development. In other words, concepts belonging to linguistics can only be attacked or defended in terms of the foundations and techniques of linguistics, while concepts belonging to physiology can only be dealt with where the special requirements of physiology have been recognized. The concepts whose "reality" Scripture denies, and the concepts which he employs in his argumentation, belong to four scientific dimensions that, though closely interrelated, still differ from each other in principle. Each of these dimensions has its own specific scientific requirements, its own object. The concept of *speech sounds* is and can only be an object of linguistics; it is the essence of a norm which has to be defined structurally by reference to geographical, sociological, and historical factors. This norm, by means of its structure and process of sound change, defines in scientific terms its geographical, social, and historical distribution and, as such, forms a prerequisite for intelligible communication on the part of a language community. In this sense, and only in this sense, do speech sounds have scientific "reality." With this concept of speech sounds are coordinated, first the psychological concept of phonetic perception and the motor aspects of speech, along with all the various prerequisite conditions as regards psychological presence and set formation; secondly, the concept of phonetic articulation, which contains requirements of the physiology of speech; and finally, the curve representing the sound in question, a curve that acquires meaning and content only within the framework of physical acoustics. From the point of view of cognitive theory, the idea that the reality of articulatory movements or speech curves should possess greater dignity than the reality of normative speech sounds is of no significance, but is merely simulated by the fact that the articulatory movements or the physical curves are visible; that is, by the close relationship, in cognitive theory, of physiology, anatomy, and physics to perception.

If Scripture is arguing his case by reference to physical and physiological factors, then it needs to be pointed out from the start that he simply *cannot* come at the linguistic concept of speech sounds by this route. The question which has to be put with respect to sound and articulatory position or sound and sound-curve is in no way a question which can decide as to the reality of any of these constructs. It merely concerns the special form of coordination of these constructs with one another. Thus questions of this sort cannot take the form: "Can the concept of speech sounds be proven, disproved, or even only modified by that of sound curves or articulatory position—and if so how?" but must rather be worded as follows: "What physiologically definable articulatory position or articulatory movement (we shall disregard this for the moment) shall be assigned to this or that linguistically defined sound, this or that speech curve, or this or that phonetic perception?" As these considerations should make clear, questions of this sort, if put in a way appropriate to their subject, fall within the ambit of the concept of a definite *system of sciences*. Scripture's arguments, however, involve the concept of a unitary science for whose purposes the methods of natural science are applied.

Scripture's objections, so far as they touch upon the concept of speech sounds, can be disposed of fairly simply by the sort of considerations we have noted above; yet as far as the concept of articulatory position is concerned, they require more painstaking examination. Above all, we need to ask how effective X-ray films have proved to date as evidence, from a physiological point of view; that is, what has X-ray film achieved or shown itself capable of achieving so far. When we look at the Gottheiner-Gutzmann films, we do see an apparently continuous movement of the various organs of the vocal tract. It would be bad methodology, however, on the basis of an impression of this sort, to regard the efforts, some of them experimental, of sixty years to assign particular articulatory positions to particular sounds as proving nothing at all. To do this, we would have at least to undertake a quantitative evaluation of the films themselves, a task which, up to the time of publication of the first edition of this book (1936), had scarcely been attempted and which, in any case, would necessitate a whole series of preliminary investigations, but without which any quantitative treatment of films of this sort cannot be entertained.[25]

In accordance with the normal procedures of the film industry, the film made by Gottheiner and Gutzmann is run at a speed of 24 frames per second; this speed is produced from Gottheiner's film, which is taken at a speed of 8 or in some cases 12 frames per second—the speed is then trebled or doubled for the purpose of projection. This alone reveals the scanty conclusive power of the optical impression we gain from the projection of the film. If our eye were a better observer than it is, we would, in fact, receive an impression of "positions" for the very reason that in each case two or three identical pictures follow each other. What actually happens, however, is that even 8 or 12 different frames per second suffice to give us an impression of movement; and investigations carried out on the physiology of sensation show that, under some circumstances, even fewer images per second are enough to convey an impression of motion.

If we establish the average duration of a sound and of an "articulatory position" at about 6/100 sec., we see that a speed of approximately 16 frames per second would be required to obtain even a single picture for each sound and each articulatory position. Since, however, the film made by Gottheiner presents at most only 12 different frames per second, the question as to whether there are such things as "articulatory positions" could not be answered by reference to this X-ray film even if we were to measure and plot the positional variations of the articulatory organs from frame to frame in the same way as it has been done in the case of graphic representation of gesticulatory movements.[26] Obviously, the vocal tract is not a machine in which the individual parts click into new static positions with great rapidity; but even a film of such a machine would give an illusory impression of continuous movement on the part of the individual components. Thus, the "doctrine" of articulatory positions, which is physiologically speaking misleading, cannot be weakened by reference to Gottheiner's X-ray film, but rather requires a deeper analysis of the relevant problems.

To achieve finality on this question it will be necessary to record a concrete conversation or a corpus of actual verbal utterances with the aid of an X-ray sound film so as to permit, first, reproduction of both the visual and the acoustic processes involved and, secondly, a quantitative analysis of both the sound curves and the photographic images. For these purposes, however, it will be technically necessary to attain a speed of up to 24 or if possible 48 or 50 frames per second. In this connection, particular stress must be laid upon ensuring that particular speech sounds can be assigned only to normative articulatory positions or movements when a given, structurally defined language is spoken in the usual manner. Physiological and physical investigations of articulatory movements or sound curves can be carried out only with reference to actually spoken material; that is to say, in respect of speech which has occurred here and now in this particular form. If then linguistics may be justified in abstracting from systematic examinations of actual pieces of concrete discourse of this sort and in proceeding directly into its dimensions of the succession of linguistic norms, preliminary stages of this sort must be included whenever questions of coordination are to be approached systematically.

These considerations require the introduction of the statistical concept of variation, within the structures of physiological investigation of articulation and of acoustic analysis of sound-curves; for in both fields the concept of a norm can be realized only by shifting over to that of a *distribution maximum*. If then the linguist wishes to return from the investigation of discourse and its unique phonetic formations to that of linguistic norms, he will be able to assign a given speech sound to the variation of physiological or acoustic phenomena.

On the basis of theoretical considerations alone, this can be said about the activity of the organs of the vocal tract: the concept of "articulatory positions" in speech cannot be used to mean more than the regularity of a rhythmic division of the relevant motion into phases, which will probably be reserved for X-ray cinematography or X-ray sound film techniques to grasp.

When we say about someone that he is speaking, we mean that the person concerned is talking to somebody about something in a particular unequivocally definable situation. What we mean is that the speaker's thoughts are directed to an object which he indicates by means of symbols which accord with certain norms. These linguistic norms of the language community, as he gives them forth by means of complicated movements of his articulatory muscles, are directed to the object of his discourse, to the speech partner and, finally, to the overall situation in which the conversation is taking place. We possess a fine ear for these manifold considerations. Even a slight neglect of one of these factors may suffice to cause a discord in communication during a conversation, so that the speaker must allow these manifold considerations constant play.

Speaking therefore means something rather more than merely giving forth speech sounds; in fact, it signifies this only to the same slight extent required for the speaker to be aware of the activity of sound production in speaking. He thinks of what he is talking about and perhaps gives some thought to the form of his question or response most appropriate to the situation; but he certainly does not think consciously about the necessity of setting particular muscles in motion in order to produce certain sounds which, arranged one after the other in a certain temporal succession, represent meaningful symbols or words which go beyond themselves and indicate an object. We know that we do not improve our discourse by deliberately thinking about how we are speaking—just as we do not improve our performance in running if ,while we are actually running, we think about precisely how we should move our legs. To give an example, one of the methods accepted fairly generally today for dealing with stuttering is that of diverting the patient's attention from the *how* of speaking to *what he is speaking about*, a process in which the situation and the conversation partner have an important part to play.

In any case we cannot simply expect that perfectly comparable conditions exist both in speaking in the sense sketched out above and in conscious production of single sounds or words. When investigating articulatory movements, we should as far as possible avoid introducing an error into the fine tuning of this play of muscles by drawing the attention of the speaker to the muscles or organic movements whose activity we wish to register for the purposes of our research. For this reason, we must give precedence to those recording procedures which allow us to investigate the speaker's articulation, if not without his knowledge, then at least so as to let him talk about something as if the *how* of his speech were not being recorded at all. Therefore, in investigating the interplay of the various organs active in speech, we should not have the subject utter individual sounds or words, but rather meaningful sentences; where possible, answers to unexpected questions. We then should either isolate the various organ movements and ascribe them to particular sounds in the completed recording or else use the conscious production of isolated sounds or the conscious movements of individual organs only as an incidental supplement to those procedures which delve less deeply into the structure of sound formation.

In the speech of everyday life—which is of course preeminently worth investigating, since it is ordinary, traditional vernacular—our attention is not even directed towards the general impulse which triggers off the complex interplay of respiratory muscles, laryngeal musculature, motor activity, vocal tract, mimicry, and gestures of trunk and limbs. And we are certainly not concerned with observing the whole range of different impulses which, in exactly the opposite way, are first deduced from the existence of the movements being observed.[27] The individual movements of the muscles do not require conscious individual impulses; and, in this regard, it should also be noted that—apart from the cerebral speech centres in which the relevant control mechanisms can be localized in a rough primary fashion—there are no organs specifically designed for speaking. All the so-called "organs of speech" retain their original vegetative functions, consisting, on the one hand, of the organs involved in breathing (the respiratory tract) and, on the other hand, of the organs which serve to take in the body's food supply (the alimentary tract). It is only during the exercise of their special functions in the act of speaking that the organs of spech enter into a loose working arrangement in which, according to circumstances, almost all the striated and smooth muscles of the body are capable of taking part.

By *articulation* in the narrower sense, we do not mean the activity of all the muscles required for or actually utilized in speech, but merely the interplay of the organs of the vocal tract—that part of the respiratory and alimentary tracts which lie above the larynx, the so-called supraglottal regions. As regards these organs, the elements which play the most important part in the process of speaking are movements of the lower jaw,[28] the tongue (including its root), the soft palate, the pharyngeal musculature, the lips, and the cheeks. No methods yet exist which allow us to register the complex interacting of all these organs without drawing the speaker's attention and thus influencing his behavior.

At this point, the concept of recording needs to be subjected to a short critique. Recording should not be understood as an imitation of nature. It would not be *recording*, and should under no circumstances be the purpose of recording operations, to attempt to imitate the speech process or, for example, the articulation of the organs of the vocal tract. Apart from the fact that any such attempt would be bound to fail, any scientific value it might have if it did succeed would not reside in its being an imitation—this would only provide us, at best, with a duplication of the process in question—but rather in that, having been achieved as a result of an insight into the regularity of the process, it would provide us with a demonstration of the accuracy and completeness of such an insight. Recording in the true sense means capturing or representing one or only a few aspects of a process—aspects which bear a concrete scientific relation to the investigation which we are conducting—in such a way that, in all cases, transposition of the actual course of events into a new, representational medium can be effected. This would bring about a state of affairs in which the result of the recording operation is not the relevant process itself, but rather a symbolic representation of it.[29] At the same time, though we cannot do more than mention this here, the notion of recording is always conditioned by that of the natural regularity of the recording instrument and of the recording procedure.

The methods of recording the activity of the organs of the vocal tract which least affect the speaker's performance are those of sound film and X-ray cinematography. Even as regards these, it does not matter very much, from a scientific point of view, that they reproduce the actual activity of the organs concerned or seem to reproduce them; what matters is that, clearly recognizing the type of transposition which they represent, they can be incorporated into scientific discourse. Scientific analysis must do justice to the principles governing their scientific usability, and here that means quantitative analysis.

When articulatory movements are recorded by means of sound film or X-ray cinematography, movements of bodies in three-dimensional space are projected onto a two-dimensional surface in such a way that, first, the third spatial dimension (seen by the observer as the dimension of depth) of both the bodies and their movements appears in the distortion of the perspective[30] and, second, a discrete number of photographic images per unit of time divides the temporal progression of a continuous movement into individual phases and fixes it in this form. By virtue of slight fatigue on the part of the eye in passing from one impression to the next, these phases, from a certain number of pictures onwards, coalesce again to give an impression of movement[31] distorted in their perspective. Here we take no account of the difference in distortion in films on the one hand and X-ray films on the other (the X-ray image is after all a type of shadow-picture!), or of the distortions which result from the limited efficiency of the registering equipment. The time continuum[32] on which the continuity of the movement in question is based can be represented in a graphic projection of the movement as caught in the individual cinematic images; it is shown by interpolating the various individual values on the basis of the space continuum which the abscissa of the coordinate scale represents.[33]

Cinematic projection of the organs of the vocal tract on a level surface leads, beyond the problem of analyzing the movement of a body in space, to the problem of quantitatively apprehending curved lines—representing the boundaries of the bodies concerned—on the two-dimensional surface of the cinematographic image. We need not, at this point, go into the necessity of reconstructing the spatial structure of the organs during the individual phases of movement from the moving lines which the shadow forms of the film or X-ray image represent, a necessity which may arise either in the interests of evaluating perspective distortion or as a result of having photographed subjects from various angles.

Measurement of these curved lines, to begin with, has only the function of relating the various stages of apparent motion fixed in the individual cinematic images to each other and to compare several independent takes of the same person and, eventually, takes of the articulatory movements of several persons in a quantitative way. Here, of course, care must be taken that the various speakers all belong to the same language community.[34]

At this stage, the numerable units of perceived articulatory sounds or of the written symbols corresponding to them must be projected into the phases of movement[35] susceptible of graphic representation. Understanding of spoken discourse requires the hearer to relate what he hears to the sound norms of the language concerned; but, if we are to analyze articulatory movements or diagrams representing them, we have to decide what has in fact been uttered or heard; that is to say, we have to take cognizance of the individual and unique way in which these norms have been made concrete in the particular case. It is, therefore, not the sound classes of a particular language that should be assigned to movement curves of this sort, but rather the numerable "units of sound perception," so far as these can be established. Projection of psychological "phases"—that is to say, perception events which do not lend themselves to precise linguistic definition—onto articulatory movement-curves will be a task of scientific relevance only in fairly rare cases. Projection of psycho-linguistic units onto articulatory curves leads to a segmentation of these curves into units which are also susceptible of enumeration—that is, so-called articulatory movement-units. Corresponding to the psychological situation, unitary complexes of movement arising from purely optical or physiological segmentation of these curves should also be called "phases." We must examine to what extent the movement units coincide with movement phases of this kind. If the numerability of the movement units is established beyond doubt by virtue of its being possible to distinguish sound classes and sound realizations, the numerability of the movement phases is still not guaranteed, because of the low degree of precision possible in the purely optical division of a curve. It is for this reason that "purely" physiological differentiation of speech sounds is not possible, and for the same reason we cannot derive the distinguishability of speech sounds from physiology. From this, we see the necessity of never registering articulatory movements without simultaneous sound recordings on tape. Only by subsequent transcription can we establish what has to be coordinated with the articulatory diagrams. Correspondingly, in the gesticulation diagrams, we have to distinguish between gesticulatory movement-units set out there and the gesticulatory movement-phases arising from purely physiological analysis.

The situation is quite similar in the case of the allegedly easily numerable letters that occur in *writing*. Certainly, in printed script, the letters can be counted without any difficulty, because, for technical reasons, they are set up individually in type. Even here, however, we must have advance knowledge: in German, for example, we cannot immediately see from the written script that "sch" represents a *single* sound, or that "ch" can stand for both the *ich*-sound and the *ach*-sound, or that in many cases "t" represents both itself and a following aspiration, so that we have a situation in which counting of sounds and of letters will lead to different results.

But what is the situation as regards *handwriting?* How, for example, can we count the letters in the handwritten word "Forschung," shown in Figure 5?

Forschung

Figure 5

We can only do so by comprehending the whole written image of the word and, on the basis of this compound entity, assigning the curves and loops of the word to its normative written form, then counting the letters in this word, and then seeking representations of these units in the curve sweep of the handwritten image. If we had not carried out the assigning operation described, we would be able to distinguish only isolated scribal phases and would lack reliable principles upon which to distinguish and count phases postulated on a purely optical basis.

Operations whereby curves are assigned to the normative written form of words are not carried out in a secondary or supplementary way by people who can read; as in the case of speaking and hearing, operations of this sort are primary. Even the subtlest analysis of writing patterns can contribute very little to interpreting the whole; it cannot replace assigning them to a known written norm. This is seen from the fact that, as a rule, we are capable not only of referring the whole manifold range of individual handwriting patterns to a normative script without difficulty (we can, so to speak, survey without effort the whole range of deviations), but also of distinguishing between identical letters accurately and almost unconsciously and of assigning them to different phonemes when required by the structure of the context. In the words illustrated in Figure 6, for example, the first letter of each second word has not been written afresh each time, but reproduced photographically, thus being, for practical purposes, absolutely identical in each case.

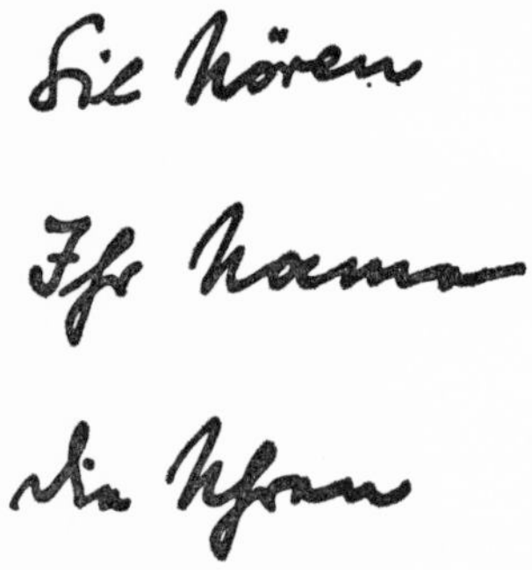

Figure 6

In spite of this, every reader assigns these identical patterns unequivocally to different written symbols. If he did not do so, written communication would break down. Very much the same thing happens in the case of the relationship between units of sound perception and the movement curves of the articulatory organs. Without a linguistic norm to refer to, patterns would be created which, though "objective," would mean next to nothing and which would exhibit different, scarcely interpretable phases. When related to linguistic norms, they explain the speech activity. It follows from this that it is not from these patterns that we can go on to set up sound classes or units of sound perception, but that the patterns are dependent upon the prior existence of the norms or classes.

To sum up: Diagrams which represent the articulatory movements of the organs of speech are divided up, by reference to psycholinguistic classification of experiences of speech perception, into numerable "units of articulatory movement." Purely physiological segmentation of these curves is carried out because of the indefinite nature of the concept of phases, in terms of "phases of articulatory movement" which are not immediately numerable.

The same is true, though it need not be discussed here a second time in detail, of the process of analyzing acoustic speech curves; that is, curves derived from recording operations, that represent the oscillation of air molecules conditioning each and every act of linguistic perception.

Certainly a segmentation of spoken words into individual sounds cannot be justified on the basis of the relevant curve forms and acoustic analysis of these curves. It can, however, be asked just how a perception event broken down into its components by linguistic analysis should be related to these curves. If, for example, the German words "es geht" have been spoken into a tape recorder, then played back and translated into curves, we are justified in searching for the physical representatives of the six sounds, including aspiration, which are involved here. If we can only discover an organization involving four or five sounds, then the curves must be investigated and analyzed or compared with other recordings until we arrive at a division to which it is possible to relate what has been perceived and, from the linguistic standpoint, divided up into its proper components psychologically. Here it can of course happen that a distinct physical organization may necessitate auditory checking and that we may discover subsequently that /t/ does not carry aspiration. Basically, however, what we derive from the curve is not that this or that *must* be heard, but rather that what has been heard and broken down into its components by linguistic analysis *must* also have its proper physical representation. In itself, the curve cannot give us any information about this in a direct way, since even sonagraphic analysis of a curve of the sort in question is incapable of showing what sort of auditory impression the oscillation under investigation *must* correspond to, quite apart from the fact that, because of phase displacements in the individual component, these factors represent a thoroughly insufficient means of sound differentiation. Analysis of this kind only gives us a *formal description* of the empirical curve, and never the auditory-linguistic description, which alone is relevant for our purposes. It would be exaggerating the capabilities of sonagraphic analysis and the Fourier integral if we were to regard them as a means of arriving at psychological classifications.

Where they can be of use is in indicating discrete elements in a classification (distinctions perhaps not visible to the eye) and their location, and curves can facilitate the process of relating these elements to auditory impressions. Not only is purely physical segmentation of a physical sound curve into linguistic segments—sounds, syllables, words, etc.—not possible; but, even in the case of psychological interpretation, greater difficulties arise than is commonly assumed, for the very good reason that perception is not the "effect" of a "stimulus," but rather experience and stimulus can be related to one another only by coordination of two different scientific points of view.

> If it is known, as regards two acoustic events, that the stimulus occurring in both coincides as to the number of oscillations and the type of wave, then it would appear that the psychological impression received cannot be described in any but the following way: the same tone, differing only in intensity. If doubts arise as to the unconditional descriptive applicability of this claim, it can be pointed out that it can be unambiguously confirmed at any point of time by any desired number of subjects. In this case, however, the psychologist is not justified, even on the basis of a statement corroborated a thousand times over, in saying, as a matter of metaphysical realism, that the tones concerned are *the same*—and here we are not dealing with the number of oscillations, but rather with the sensations which have been produced. We continually overlook the fact that the judgements of human subjects are almost always made on the basis of a functional training process and thus represent an *application* of learned functions; this is one of the inherent evils of our experimental activity, with all its pretensions to exactitude. In the case which we are dealing with, this means that the subjects have *learned* to put tone impressions which may be essentially and in point of origin quite different from one another into the same bag and to treat them as if they were identical. (We use this metaphorical and neutral expression in order to avoid the concept of abstraction, that piece of Aristotelian logicism which is continually being dragged in unnoticed by Psychology, and which stands in such need of clarification.) [36]

At the most it can be stated by analogy, and solely on the basis of earlier relevant investigations, that an a-type or an o-type auditory impression will probably correspond to this or that curve pattern or this or that combination of formants.

Here again, purely physical "curve phases" have to be distinguished from the segmentation of a curve into numerable "units of pitch" which are brought into existence in only a secondary way as a result of psycholinguistic segmentation of a word into speech sounds. And we still have to ask to what extent the curve units coincide with the curve phases; that is to say, we have to ask to what extent the units can be coordinated with the curve phases such as can be established by purely physical means.

In this connection, we need to look once again briefly at the considerations raised earlier in connection with the segmentation of two adjacent curve units within a continuous curve pattern: the line of demarcation between them cannot be a clear and simple one, since, in a curve capable of being enlarged to any desired scale, how could a single, exact point of demarcation be defined? The sound curve representing the motion of oscillating air particles must be continuous, because these air particles, whether they actually move or not, have duration in time. On the other hand, any segmentation into numerable soundcurve units has to be discontinuous. This is involved in the whole concept of coordinating the psycholinguistic dimension with the physical one. From this, it follows that the curve pattern must exhibit transitions between each pair of units and that, in the sound curve in this connection, we must distinguish central, nuclear elements from transitional, linking ones.[37]

The central phonometric problem of *coordination* is thus, in the acoustic and physiological dimension, a problem of relating continuous phases to discontinuous perceptual units which are divided up, however, not on a psychological but rather on a linguistic principle.

IV. Phonometric View of the Sound System

It has been shown in the preceding sections that the two concepts of set theory—the ability to distinguish and the ability to segment—are among the assumptions that a linguist must make in order to be able to analyze a linguistic structure and to relate it to other linguistic structures.

No attention was paid to the question of whether there are different forms of distinction and distribution of linguistic segments within each language, and whether perhaps that must be so to make linguistic communication possible. This would mean that not only the ability to be distinguished and to be segmented, but a *system* of distinctions and a *system* of distribution of linguistic segments, are among the conditions of linguistic analysis and thereby among the presuppositions of every linguistic structure.

It should make no difference by which route the individual linguist arrives at the knowledge of such systems. In seeking empirically in one language a multiplicity of forms of distinction or of distribution of linguistic segments, he must know that to seek these is inherent in the concept of linguistics itself. The possibility of systematizing the differences and forms of distribution so found belongs to the conditions for linguistic analysis as much as the two categories already discussed: ability to distinguish and ability to segment.

1. SYSTEM OF DISTINCTIONS AND OPPOSITIONS

Ever since its beginning, linguistics has conceived its object exclusively to be the system of languages in the stricter and real sense of the word; that is, with relation to human speech. At the same time linguistics has made allowance for the fact that all "higher" systems in which the system of languages could perhaps be placed—the system of symbols, the system of cultures, the system of organisms, etc.—are not definable without presupposing language in the real sense of the word.

From the outset linguistics has thus regarded itself as an autonomous science in the system of sciences, and understood the particular language as the structure of its realizations valid for a community in a geographic-social-historical area. For this reason these structures of individual languages, as von der Gabelentz[1] calls them, were and are analyzable approximately, if the realization of their forms is not taken into account. Here the systematic character of the variety of languages is guaranteed not so much by quasi-genealogical relations of individual, related groups of languages, but rather by the relative openness of the structure of each language as against the structures of all the others. This is true with regard to geographical, social, and historical notions. Foreign words and slang, which can break through "closed" systems—with respect to the phoneme inventory also—are only two particularly obvious aspects of this situation.

Every linguistic judgment presupposes not only segmentation and distribution of segments, particularly of the phonemes, but for each particular language a *system* of distinctions and distributions. To look for such systems and to discover their relations is among the legitimate tasks of linguistics, and to accomplish these it can no longer forego the use of recordings and their auditory and quantitative analysis.

Whenever the system of transmitted languages is treated here as *one* form of symbolic languages, no attention is paid to the fact that this is only partly possible because the system of "real" languages is logically prior to the system of symbolic languages (including "animal languages"), since the "real" languages are among the conditions for the scientific definition of all the other types of languages.

If with the Morse Code one disregards the question of necessary intervals (that is, segmentation), then its construction is based on two principles of distinction: on a binary opposition and on combinations of these two contrasting elements. (Combinations of from one to four elements are allowed for the letters of the Morse Code.) The principle of forming combinations is here secondary as it presupposes the relationship of mutual exclusion. If one were to attempt to form combinations without regarding the binary character, by combining dashes of various lengths, one would quickly reach the limits of distinction and unequivocal technical realization of the elements. So the variety of the inventory of a given language relies on these two principles: the phonemic system and the distribution.

If one were to measure the duration of given symbols in order to "understand," which means here to distinguish and to distribute them, then this principle could be extended at will: to 1/10 sec., 2/10 sec., etc.; or 1 mm, 2 mm, etc. Very strict limits, however, would be set to such extension if one had to get by without measurement; that is, to restrict communication to auditory and visual perception. If texts given are to be heard or read, the connection of the two principles of distinction cannot be avoided.

What is expounded herein on the system of the Morse Code is also valid for language spoken or heard.

In many languages, the distinction between voiced and voiceless obstruents is phonemic. High German distinguishes for example voiced /b/ and voiceless /p/, or more generally: voiced media /b d g/ and voiceless tenuis /p t k/. In this context we will disregard the problem of their different degrees of tension and aspiration. In initial position, High German distinguishes (for example) 'Bann' and 'Panne,' 'Birne' and 'Pirna,' 'dann' and 'Tanne,' 'gern' and 'Kern.' These examples show not only the opposition between voiced media and voiceless tenuis, but they also demonstrate (disregarding 'gern' and 'Kern') that all the words compared here display still other differences than the initial consonants; furthermore they exhibit the extraordinary variety of spheres of meaning from which these contrastive word pairs derive. This means that little importance needs to be placed on the realization of the opposition voiced-voiceless, if it were only a matter of avoiding confusion between words. Regarding the homonyms of High German, we know how seldom they disrupt communication. Their existence disturbs us so little, that numerous homonyms are scarcely familiar to us as such. Many speakers will have scarcely ever noticed that we employ the same word-form for the Atlas Mountains, for a collection of terrestrial and celestial maps, for the topmost cervical vertebra, and for a textile. If in spite of this we place importance on the distinction between voiced and voiceless phonemes, then other factors must play a greater part than the endeavour to avoid misunderstanding.

For these reasons the opposition of voiced and voiceless in High German is in fact much less preserved in speaking than our pronouncing dictionaries demand. Generally moreover, they describe less the actual than the prescriptive usage, usually without sufficient regard to actual usage and too much under the control of the (printed and written) letters with their different transitions and possibilities of confusion. Measurements of voicing show all shades between complete voicing and complete lack of voicing, with marked overlapping of the variations of both classes demanded in the pronouncing dictionaries. These transitions form a physiological continuum inasmuch as, between each two realizations of limited voicing or lack of voicing, a further transition is conceivable. This physiological continuum could theoretically be developed into the basis of a polynomial system of symbols—just like the time continuum, which guarantees the differences in the realization between the long and short symbols of the Morse Code. But it would have the same limits of perception and production or articulation as a Morse Code, which only knows differentiation between longer and shorter symbols.

Naturally the same is true of anatomical continuums of space and physiological continuums of tension: the High German *ach-* and *ich*-sounds are velar and palatal fricatives. Both strictures are situated on the space continuum of the palate, which extends from the ridge of the upper incisors to the uvula. Rounding and unrounding of the lips show a similar continuum. Even the different states of tension of the individual muscles which take part in speech show such physiological continuums, which could theoretically be divided into any number of measurable but not perceptible units. None of these one-dimensional physiological systems would be adequate for developing a system of symbols which could secure and express the richness of language. Because of the physiological circumstances of the speech act, various physiological articulation-types *must* consequently be combined in order to obtain a phoneme inventory rich enough to record the vocabulary of a language—or even only the vocabulary of linguistics. For these reasons the linguist, with every language, is obliged to seek its *system of combining* physiological characteristics connected by continuums. That it is generally a matter of binary systems and thus of *oppositions*, as with the Morse Code, is due to the fact that the elements of binary systems are easiest to distinguish on the levels of articulation and of perception. Such a distinction, however, need not always be realized in particular.

It was mentioned above that the reason for the desire to realize phonemic oppositions in the context of spoken language is by no means always—probably even quite seldom—to be sought in the endeavor to avoid misunderstanding. Generally the situation and the context take care of this so thoroughly that one could probably dispense with the realization of many oppositions. There are other categories—nonlinguistic ones, and therefore little regarded within the framework of linguistics—which here play a far greater part.

Oppositio means literally resistance, disobedience, contradiction, opposing party (usually that party which opposes the government as "opposition"); and only in a figurative sense does *oppositio* mean the contrast of objective circumstances. This meaning employed by phonology could thus almost be called a metaphor.[2]

It has already been mentioned that such oppositions—for example, voiced and voiceless quality—are only very rarely employed in spoken language for distinguishing two sentences and hence two words similar with respect to contrasting terms. Jespersen has pointed out to what a small extent this is the case, particularly in German, by means of examples from English, French, and German: "The number of such minimal pairs is much smaller in German than in French and particularly than in English; and according to my conception of 'sound laws' it is essentially to be attributed to this fact, that in so many parts of Germany the distinction between media and tenuis has been obliterated."[3]

Such linguistic oppositions acquire their importance for the concept of the phoneme only in the context of a system of sounds released from the flow of speech. How important their part is appears perhaps most clearly from the fact that Trubetzkoy altered the sequence of the definitions of phoneme and opposition in his "Anleitung"[4] and "Grundzüge,"[5] and gave precedence to the definition of opposition in his later work. In spoken language, other forms of opposition, essentially closer to the original meaning of the word, play a much greater part than the desire for differentiation by means of "contrasting" sounds in minimal pairs.

If one asks a German-speaker why he says "Paul"—there is no word "Baul" in German—he will answer: because that is what the person in question is called; or, because this name is pronounced so. That is, he speaks as he does because this corresponds to the current (traditional) usage, and because he too wishes to follow this usage, not to change it. Here a tendency towards *conservatism* contrasts with a tendency towards change. The fact that languages only change slowly and generally without the speakers' noticing was probably first pointed out by Dante.[6] It is based principally on the preponderance of such tendencies towards conservatism as against tendencies towards change. Rational motives of easier communication probably play a considerably lesser role in this trend than other deeper and more far-reaching tendencies. Outside language, too, such conservative tendencies, which cannot be expressed rationally, are operative—for example, in the field of folklore.

The field of fashion shows that tendencies towards change can *contrast* with such conservative tendencies. However, there are such tendencies towards change in the linguistic sector also. One such tendency towards change is fed by scientific and technical progress, which must continually "introduce" new words, by which traditional words can be rendered obsolete. Here mutually opposed views of persons and groups of persons "contrast" in the original sense of *oppositio*.

Geographical and social oppositions show that such oppositions can exist in the phonetic field also. To the question why a speaker says "Paul"—with an aspirated p^h in the usual High German pronunciation—there could come the answer: because I don't want to speak Saxon; and an opposition both geographical and social could be concealed behind this answer. Many forms of "refined," precious, or pretentious speech belong here. Generally a linguistic opposition deriving from social pretension or contrast can then be indicated; for example, when someone says "müd" instead of "müde," or by analogy with English "weekend" says "Wochenend" instead of "Wochenende."

Only in the framework of a system of distinctions yet to be developed and of a system of oppositions yet to be projected can the phonological opposition find its place.

2. SYSTEM OF SEGMENTS AND OF VARIANTS

So far, the problem of segmentation has been treated only with respect to the smallest linguistic unit, the sound class. It was defined as an auditory unit, presupposing that only someone knowing or even better speaking the language in question acts as a listener. By this condition or requirement of phonometrics, account is taken of the precedence of linguistics over the "purely" psychological aspect, as regards both perception and phonation. This requirement derives from the presupposition of linguistic research that speaker and listener can only communicate when what they say and what they hear—speaker and listener are basically interchangeable—are subordinated to a common code, which is handed down and is valid for the whole speech-community. Here the problems of bilingualism and multilingualism, of two and of many layers, and of the transitions from one language, dialect, or speech-layer to the other may be disregarded.

This primacy of the linguistic aspect for audition and phonation means, on the basis of the *traditional* character of this code, a limitation on the distinction between *diachronic* and *synchronic* aspect introduced by Saussure,[1] and a peculiar amalgamation of both viewpoints, which von der Gabelentz is aiming for when he attempts to show "how both must finally be interwoven." [2]

When von der Gabelentz speaks of "equality" or Saussure demands precedence for the synchronic viewpoint over the diachronic, both can be only partly valid. In fact, at the beginning of every linguistic analysis of a *living* language should stand a synchronic presentation of its structure, by which in turn diachronic research is promoted. However, this structure can only be understood as a code valid by its transmission, which in detail has the effect that, for example, the definition of the smallest linguistic unit cannot be made without reference to this transmission—and in a double sense: the listener understands by referring what he hears to transmitted norms. This hearing in its turn reacts on these categories of the listener's understanding, by either confirming or altering them. The speaker, too, speaks within transmitted norms and can transform these norms by this way of realizing them. Here we will not consider whether such transformations via speaking or listening occur consciously, or whether they can be consciously made or not. In connection with the statistical structure of realization—without which there "are" no linguistic structures at all—one must consider that certain alterations in these structures can be ascertained and defined only by statistical procedures. In this case, the speaker or listener cannot be aware of them, because both have to restrict their statements to the "presence of the production" [3] of speaking and listening, while these types of statistical "forces" can only be "at work" and demonstrated in longer texts.[4]

A closer examination of the relations between the auditory-synchronic and linguistic aspects compels us to examine the relation of *sound class* and *allophone* or of *combinatory variant* of the phoneme.[5]

Three examples may further the discussion: the ich- and ach-sounds are in Modern High German two allophonic variants of the same phoneme, for reasons which can here be assumed to be known. For other reasons, the /k/ in "Karl" and the /k/ in "Kind" are two allophones of the same phoneme. And finally, the aspirated /p/ at the beginning of a word and the unaspirated /p/ after /s/ or [ʃ] are similarly two allophones of the same phoneme.

However, these three pairs of allophones—in the position in which they are to be expected and required—by no means need to be always spoken in the same perceptible form. This is so not only for geographical, social, or contextual reasons—that is, reasons which could be cited and therefore used for establishing classes—but also for non-verifiable reasons that are inherent in the realization of the position in question. In this sense, these articulations are "purely fortuitous," and therefore their form is not foreseeable and their extent not predictable.

When a number of trained listeners, familiar with the language in question, can agree on a phonetic value and a phonetic symbol for the realization of the allophone in question, we call such a variant of an allophone a *speech variant.*

In practice it quite often happens with transcription that even experienced listeners, even when they all speak the language being treated as their mother-tongue, cannot agree on a phonetic value and a phonetic symbol for the realization of the allophone expected in the position under discussion. In such cases—after suitable tests, by which errors through lack of concentration are eliminated—we speak of *transcription variants*, which in phonometric investigations are considered necessary to be listed in the text for reasons of later statistical analysis of the results of measurement.

Even in the cases where all listeners agree in their judgment regarding all allophone or a speech variant, observation and measurement by means of oscillograms yield varying values. One must assume that between every two such values a third is always conceivable; for this reason we spoke above of a continuum in a sound class. In this sense the oscillogram (the oscillation of the air-molecules made visible) allows in principle an individual determination of every realization of every sound class.

Since a definition of this individual form of realization is only possible by measurement and since measurement, as demonstrated above, presupposes a scale with finite gradation—for example, 1/100 sec., quarter-tones, millimeters, and so forth—this definition is again that of a class, which can be burdened with a number of individual realizations no longer distinguishable by means of this scale. We call these classes *phonometric variants.*

These variants—combinatory (allophonic) variants, speech variants (speech sound classes), transcription variants, and phonometric variants—are all part of the conditions for linguistic description. Their variety raises the question, whether *a system of all conceivable variants* is to be required, in which also the distinctive (semantically differentiating) opposition would be understood as a variant. In seeking such a system, the concept of variants would have to be broadened: already the measurement of melodic or dynamic alterations compels us to step from speech variants, allophones, or phonemes to larger groups of such units. In such a system, *phonemes, prosodemes, morphemes,* and *syntactic units* would have to be distinguished, since all these segmental forms are expected to vary in their realizations.

These tasks can only be accomplished with verifiable exactitude on a much larger corpus. Such a corpus was created by the *Deutsches Spracharchiv* between 1955 and 1965; it contains some 8,000 tapes of literary and colloquial German and German dialects. At present, their processing and electronic storage is being undertaken for comprehensive linguistic evaluation.

V. Execution of Phonometric Procedures

The problem of writing texts begins with the determination and expression of the inventory of the sounds of a language. In other words, one of the first problems one encounters is that the criteria according to which the different sounds are determined are not uniform in the current phonetic alphabets such as that of the IPA. Phonological, auditory, acoustic, and articulatory criteria are used side by side and often in a hodge-podge fashion. It is desirable that criteria should be available at all three levels which would recognize the preeminence of the phonological principle and would take into account the necessity of coordination with phonological parameters. As long as such perfected systems do not exist, one must manage with one of the present systems such as the IPA system, which we will utilize here.

1. COMPILATION OF TRANSCRIBED TEXTS AND PHONOMETRIC TEXT LISTS

After recording the spoken language on recording devices which are adequate for the linguistic purpose in hand (phonograph records or tape recorders,[1] sound films, or X-ray films with a sound track), the first requirement of our phonometric task will be to write serviceable *texts*; that is, texts which are adequate for the task in hand. Phonetic investigations, and in particular quantitative ones, carried out without such texts contradict the concept of phonetics, and in particular the concept of phonometrics. They can have at most the value of providing an orientation in the investigation of a linguistic problem.

We must here distinguish three kinds of texts:

Phonemic (or allophonic) texts.

Transcriptions (or texts giving phonetic sound classes) with articulatory and auditory variants.

Phonometric text-lists.

The mutual relationship of these three kinds of texts is not easy to grasp. Although the first task must always consist in preparing transcriptions, or texts giving the phonetic sound-classes, an appropriate transcription must be guided by principles which, if formulated explicitly, would amount to the listing of the phonemic inventory and its distribution for the language or dialect concerned. In the practice of writing transcriptions this means in the first place that the listener *knows* the language or dialect of which a transcription has to be made, so that he will be able to understand what the speaker means; and in the second place, it means that the listener is able to write down what is heard by means of discrete *phonetic symbols*. The system of phonetic symbols should be such that each symbol represents one sound and each sound corresponds to one symbol. Seen in such a manner, the problem is shifted towards the definition of speech sounds or—as will be stated in more detail when we deal with (phonometric) text-lists— the definition of the sound classes.

But even the first requirement is not undisputed and must therefore be further substantiated. It must in fact be defended in two directions: against the school of "experimental phonetics" and against the *phonology* of the Prague School. Both agree—although they use different arguments—that it is not he who knows the language who is the suitable and therefore desirable person to prepare a transcription, but on the contrary that the transcription should be made by someone who does not know the language.

As far as the experimental phoneticians are concerned, this requirement is understandable although, if considered more closely, not necessary. It is therefore not shared by some of the followers of "experimental phonetics" such as Scripture. The experimental phoneticians are, or were, of the opinion that someone who does not know the language can be a more objective listener than somebody who knows the language. In their opinion, the connoisseur would know the language too well and would project this knowledge into his transcription, whereas his task is merely to describe "objectively" what he "really" observes.

Two conceptions of *objectivity* and *reality* are confused in this demand of the school of experimental phonetics. One refers to the objectivity of physics and physiology or the reality of animate and inanimate nature (which in fact do not coincide and the relations of which we do not need to investigate here). The other refers to the objectivity of linguistics or the reality of language. However we define these latter concepts, it is certain that they do not coincide with the objectivity of the natural sciences or with the reality of nature. This is shown—even before any epistemological investigations—by the history of the sciences in the 19th and 20th centuries: by the history of physics and physiology on the one hand, by the history of linguistics on the other.

Therefore a phonetician such as E. W. Scripture consistently denies linguistics the right to be an autonomous science; as he also denies to "language" the status of an autonomous reality—without being able to replace the "illusory concepts" of that doubtful discipline of linguistics by the new "concepts of reality," which he hoped to obtain from the investigations of the sound-pressure curves conducted exclusively along the lines of the natural sciences.

Trubetzkoy also prefers as a suitable listener a person who does not know the language to a person who knows the language.[2] Decisive for him—outside of his linguistic-phonological analyses—was a twofold division into linguistic phonology (at the level of grammar) and phonetics, which belonged to the realm of the natural sciences.[3] This twofold division was patterned on the allegedly objective division of the *globus intellectualis* into the natural sciences and the humanities. He considered the business of listening to be exclusively the task of phonetics and hence allocated it to the side pertaining to the natural sciences in that double science linguistics-phonetics; consequently he had to demand an "objective" transcription technique, based on the concept of the objectivity of the natural sciences.

It has been proved, however, that a person who does not know the language is unable to make transcriptions that do justice to the spoken word. His hearing is less "objective" than the hearing of a person who knows the language in question; he hears—at least in the beginning— with his own system of phonemes instead of with a system adapted to the language being spoken. This is because, without an inventory of sounds (phonemes) he is unable to accomplish his task: the writing of a text.

The alternative therefore, if we want written texts, is to renounce this demand. That such a renunciation implies also the renunciation of the adequate counting of the curves of sound pressure or the curves registering the physiological acts of speech has been pointed out above. And this is the real concern of phonometrics, in contrast to the expectations on which the so-called experimental phonetics school was based. The requirement that only those with a knowledge of the language should be listeners has in practice to be taken with a grain of salt. There are cases in which someone who does not know the language hears features of a language which will be likely to escape persons who know the language. Two possibilities must here be distinguished. It is, for instance, well known that speakers of related dialects sometimes notice peculiarities of a dialect which escape the speakers of this dialect. Speakers of German who are not from Saxony will hear the Saxon "singing" or the lack of distinction between aspirated voiceless stops and unaspirated voiced stops. Normally the Saxon speaker will not notice this. Similarly inhabitants of Zürich will hear peculiarities of the dialect of Basel which will escape people of Basel. But in these cases we are not dealing with persons who do not know the language, in the strict sense of the word, but with speakers of related dialects. Here the speaker of the related dialect understands the other dialect, relates it to his own frame of reference, and measures it accordingly. We are therefore dealing here with a special case that can lead to the clarification of auditory facts. But there are still other cases in which persons who do not know the language are better listeners than those who know the language. In modern standard German, "voiceless" stops are often pronounced voiced while "voiced" stops are often pronounced unvoiced. Strictly speaking, a mistake has already crept into this statement: the confusion of linguistic description and pedagogical rules of pronunciation. A situation in which voiceless sounds are pronounced voiced is foreign to a linguistic description. A phonetic transcription can only ascertain what *is* and not what *should be*. A carelessness in the realization of sound classes, which is not accompanied by changes in meaning or expression, coincides on the auditory side with an uncertainty in the discrimination of distinctive features. It is therefore true that persons who do not know a language are sometimes better qualified to make an auditory discrimination than the native speakers. In the example just mentioned, a Japanese would be better equipped to discriminate between voiced and unvoiced stops because in his native language the interchange is not possible as it would cause a change in meaning. Although we have to do here with people who do not know the language, they have, as a result of a command of their own language, acquired a particularly good discrimination of this distinction.

The problem hence reduces, as we mentioned earlier, to the question of what—in the context of writing transcriptions—we mean by *speech sound*. Our answer is that a speech sound is simply a segment of a spoken utterance, which a native speaker of the language discriminates from the preceding as well as from the following segment of its context and relates to the sound inventory handed down by tradition. We see that a diachronic element enters even into this stage of synchronic description.

If one lets several trained native listeners make a phonetic transcription of a linguistically and technically faultless tape recording, differences will be found between the resulting texts, which cannot be eliminated even after repeated transcriptions.

As it is not feasible to create a new phonetic symbol for each of these perceptual differences, the only solution in such cases is to add to the phonetic transcriptions the different *auditory variants* produced by the listeners. By thus adding the different ways in which the text is apprehended by the hearers—who represent in a certain sense the speech community—the social nature of the spoken language is brought into focus. The social nature of the spoken word differs from the social nature of written or printed texts as a result of the different modes of transmission involved.

When the hearers are in agreement about a sound that has been heard in a particular context and about its relation to the inventory of sound classes and sound symbols of the language under consideration, this sound will be represented by the appropriate phonetic symbol, irrespective of the phoneme[4] or combinatory allophone which would be expected. Thus the transcription can simultaneously be considered as a *phonetic text*. The discrepancies with the expected phonemic-allophonic sound symbols will be listed as *speech variants*, which will characterize a particular speaker as well as the language concerned as a whole or its different linguistic strata.

This *phonetic text* must be distinguished from a *phonemic-allophonic text*, which provides the following information: In a certain situation, depending upon the linguistic context and the dialect's geographical area, a particular speech sound will be used; it is expressed by its appropriate sound symbol. Such a sound is expected to occur at a particular place of the text to be transcribed. Such a text is called a *phonemic-allophonic text* because each phoneme or combinatory allophone of the language concerned is expressed by one sound symbol. It uses, for instance, for the two combinatory allophones of the New High German phoneme /x/ the two sound symbols [x] and [ç].

From the point of view of phonometrics, each of these two kinds of text is used for a different scientific task. The scientific rationale of writing phonemic-allophonic texts consists, in the first place, in ascer-

taining the *inventory of sound classes* of a given language community—
that is, in determining the nature and number of the sound classes which
must be necessarily distinguished in the language in question. In the
second place, the task served by such texts consists in determining the
usual pronunciation of the language under consideration. (Consequently such phonemic texts determine the range of a linguistic habit
geographically, situationally, socially, and historically.) Finally, such
phonemic-allophonic texts have a certain value for phonometrics because they help to determine the specific task of the investigations and
to define the statistical classes which are relevant in a given situation.

Also, in the writing of allophonic texts, we have to rely on a limited
number of speakers whom we can allow only a limited time to speak,
so that a finite number of words occurs in a finite number of repetitions. The pronunciation of these speakers is recorded by means of a
suitable system of transcription. But the purpose of such transcriptions
is not to characterize the pronunciation of speaker X during the recording Y and to distinguish this pronunciation from that of another speaker
during another recording. The purpose is to record and analyze the
usual pronunciation of a linguistic community of a given size. Only
what is usual is preserved, has currency in a language community in a
certain geographical and social area during a certain period, and is
transmitted from one generation to the next. This phenomenon can
become the object of linguistic investigations.

Even an allophonic text recording the pronunciation of a *single*
speaker can establish his usual pronunciation. The procedure hereby
followed, which is inevitable as well as justified, is to let the speaker
repeat separate words. We disregard for the moment the mistakes to
which such a repetition of single words can give rise. (The magnitude
of such mistakes can be easily seen when we try to pronounce a familiar word of our native tongue about ten times. By such a repetition we
do not, in fact, increase the certainty we have regarding our own pronunciation; on the contrary, we can finally become so uncertain that
at the end we are unable to give any information at all about our own
pronunciation.) Texts written on the basis of this procedure will hardly
ever provide reliable information on a pronunciation which is in itself
already fluctuating. We will therefore from now on designate with the
term phonemic-allophonic texts only those texts that attempt to render
the pronunciation of a certain linguistic community during a certain
period of its existence.

There are two requirements for the reading of a phonemic-allophonic text: in the first place the text itself, which symbolizes the usual pronunciation of current words; and in the second place, the system of sound symbols with the necessary indication of the sound value of each symbol. These sound values should be illustrated by examples. Besides this there is the task of writing transcriptions; that is to say, the recording by means of symbols of what has been said once at a specified moment.

Phonograph records and tapes have influenced our transcription procedure. They allow the repetition of continuous speech, which is unique by its very nature. And thus they permit a repeated listening to this discourse by several persons. In this way the concept of a language can be established in a systematic way.

In the writing of a phonemic-allophonic text, the question which the investigator has always to bear in mind is: what is the *usual* way in which this particular segment is pronounced by the particular informant or by a particular speech community? In the writing of a phonetic transcription, however, the leading question will be: under which of the *speech-sound classes* characteristic of the language under consideration must the speech sound be categorized that occurs in a particular place on the phonograph record or tape? This question presupposes a whole series of conditions that will be enumerated below; the different tasks and aims of phonometrics and of experimental phonetics rest on these assumptions.

The question presupposes in the first place that we are dealing in the case under investigation with *language*. The question is therefore not: do the sounds and noises of this record or tape represent language or not? The question thus phrased can be answered by yes or by no and is not a scientific question. It would only be a scientific question if a scientific method would succeed in deciding, on the basis of an immanent and non-linguistic analysis of the curves or other vibrations, whether these curves *must* represent human speech and a particular language. This idea, if consistently applied, denies the linguistic nature of language and confuses the possibility of *constructing* speech sounds with the task of describing them. These conditions remain basically different, in spite of the fact that in certain languages it is not possible to decide without further investigation whether we are dealing with speech or singing. Carl Meinhof once raised this objection during a discussion. The distinction can be made in principle because song represents a *mode of tradition* which is distinct from speech.

These considerations indicate the further suppositions underlying the question formulated above. This question assumes in fact that we are dealing with a *particular* language. This, too, cannot be determined on the basis of the quantitative analysis of curves. Relying on auditory impressions to resolve this question—that is, a hearer asks himself: "Which of the languages I know is spoken here?"—involves in a somewhat tentative way precisely what we have done in the question formulated above: the hearer assumes that he is not dealing with some artificial construction but with a real spoken language. In other words, he assumes that the speaker uses the medium of a language community. And it is only on the basis of this assumption that he approaches the task of transcription.

However, when isolated sounds or meaningless syllables have been recorded and it cannot be determined which language they are in, it also cannot be decided which transcription system is most suitable, unless we operate on the assumption that the meaningless sounds and syllables were pronounced in the *manner* of a historically transmitted language. However, as long as language and meaning of a text have not been established, there are no criteria for deciding what is normative and what is accidental in the sound under consideration. This cannot be determined by the observation of isolated instances but only by comparison. And behind such comparison stands as final criterion the meaning of the contextual situation which also serves as cynosure in learning a language. The several systems of phonetic symbols which are utilized in phonetics have therefore their practical utility: they provide a technique for writing and printing. It is necessary to be aware of the fact that the sound value of a particular symbol will be somewhat different in the various languages for which it is utilized. This "somewhat" often constitutes precisely the characteristic difference between the languages in question. This can be easily demonstrated by giving a phonetic text in a certain language to read to a speaker who knows the sound values of the alphabet used in his own language but not in the language of the given text.

The third assumption on which the phonometric question rests is the *linguistic determination of the appropriate sound-system*. We have seen that the linguist—and in fact also the speaker and hearer—must consciously disregard certain distinctions. He must, for instance, disregard the differences between the speech organs of different individuals and the inevitable variation in pronunciation if he intends to approach language from the point of view of linguistics. But then he must also have the same certainty as the speakers and hearers have as to which of the differences are relevant and which are not. It is precisely this that is meant by knowing a language and basing the transcription of a text on this knowledge. Therefore, the question cannot be, which sound do we have here at this particular point? but—as we have pointed out above—which sound of the sound system, adequate for this language, do we have here?

For various technical and linguistic reasons, it often happens that this question cannot be decided, as in the case where some words or syllables have been pronounced so indistinctly that what has been pronounced cannot be heard exactly and we can only guess from the context what has been said or should have been said. In such cases we do not speak of *units of phonetic perception* but of *phases of phonetic perception*. These phases are certainly not neglected during the further stages of the investigations; on the contrary, it remains constantly necessary to ask again and again what they may have in common; that is, whether they cannot be segmented linguistically in one way or another.

The experimental phoneticians believed that the process of hearing as well as the process of transcribing falsified the object of study or at least treated it inappropriately. The aim of the present considerations has been to invalidate this claim and to show that the very opposite is true. And further investigation will prove that measurements which are adapted to the object studied are in fact only possible if the procedure suggested here and used by us for several decades is followed.

When the hearer listens to a tape recording, he must at each sound again ask himself the question: Into which of the sound classes provided by the phoneme inventory would this particular speech sound fall? Because speech sounds are produced by human organs, we are sure to encounter variations in their production, and we must assume from the outset that, in some cases, it will be difficult to decide to which of two sound classes a certain sound must be attributed. There must be cases intermediate between two such sound classes and such cases can be expected to have particular linguistic interest.

For this reason several (at least three) persons should listen to a tape recording and each should write his transcription independently of the others. In such a transcription, only the second classes which result from the listening process will be noted down. This is done by determining the place of the particular sound in the system of sound classes or in the phonetic alphabet that is at the disposal of the listener. At this stage, it is convenient to disregard stress, pitch, length, and pauses—at least, this was done in transcribing German dialect texts.

When three persons have listened to the tape recording, the resulting transcriptions are compared, and those words, of which a part has been heard or interpreted by one person in a different way from the others, are listed in orthographic script. This is the so-called "provisional list of variants." The list is multiplied and each of the three hearers listens once more to the words on the list and is asked to transcribe them again without reference to his earlier decisions. The same procedure is repeated on a new list a few days later. In view of the size of the texts and the time interval with which the repetitions take place, it is practically impossible that the hearers still remember their own earlier decisions in each case. As a result of this procedure, we will have either three *identical* decisions, which are independent of each other, or—in the case of segments which were at first differently transcribed—nine mutually independent decisions.

On the basis of this list, the so-called "definitive list of variants" is finally produced. The list that follows refers to a German tape recording which was transcribed by three persons.

1.	eʊstn	7	erstn	1
			eʊztn	1
2.	fraithak	8	fraitak	1
3.	fraithak	8	fraitakh	1
4.	ɔkthobʊ	7	ɔktobʊ	2
5.	bəgɪn də	6	bəgɪnt də	2
			bəgɪn tə	1
6.	pro...	6	bro...	3
7.	probyʊ	7	probyə	2
8.	markt	8	mark	1
9.	axtagə	6	axthagə	3
10.	dauʊ	5	dauʊt	4
11.	das	8	daz	1
12.	gantsən	5	gansn	3
			gantsn	1
13.	hauzə	8	hausə	1
14.	gəʃlaxtət	7	gəʃlaxthət	2
15.	unt	8	un	1
16.	gəbakhən	7	gəbakən	2
17.	dizəm	8	dizm	1
18.	thagə	7	tagə	2
19.	myʊn	5	myʊən	4
20.	noiə	6	noiən	3
21.	vɪntər	4	vɪntʊ	3
			vɪntə	2
22.	fɛrtɪk	6	fɛrtɪç	3
23.	zain	6	sain	3
24.	eʊst...	7	erst...	2
25.	eʊstn	8	eʊstən	1

The numbers in the first column refer to numbers in the text; they are followed by the phonetic transcription of those words which were interpreted differently. The most frequent transcription form comes first, followed by the number of times this transcription was given by the hearers. The following column gives the deviant transcriptions with the number of times they occurred. The sum of the decisions must always be nine.

Originally this list of variants was only intended merely to list those parts of the text that had been transcribed differently and that therefore had to be excluded as doubtful from further investigations. But later it turned out that these lists of variants have a linguistic significance of their own: This became especially clear after it had been shown that the different *hearing version* (the transcriptions based on auditory criteria) varied basically in the same way as the *speaking versions* (the speech of the same persons pronouncing the same text).

This list of variants is used to produce the so-called "phonetic text," which gives the majority decision[5] for each instance of a deviant transcription. In this way, all mistakes resulting from lack of attention are excluded while at the same time the list of deviant transcriptions shows how strongly the decision in favor of a certain variant is supported. The more support a transcription finds, the more certain will be the result.

The phonetic text is from the outset written in a certain arrangement, the reasons for which will be set out in a later work.[6] This arrangement sets up so-called "blocks" and "groups" in such a way that every 100 successive sound symbols form a block. Ten blocks constitute a "group." This arrangement, which was originally used for frequency statistics, has proved to be effective also for the preparation of texts that are subjected to more recent methods of data processing.

By way of example, we will here briefly touch upon one difference in transcription and speech that occurs particularly often: the aspiration of voiceless stops. The question of whether aspiration occurs at a certain place is important for the transcription, because it is impossible to write a text without making a decision on this question for every plosive. Such decisions are also relevant to the question of what is the usual pronunciation and to the problem of its phonemic interpretation.

Theoretically there are four ways in which such a question can be decided. We could examine the oscillogram that results from the vibration of air molecules; secondly, we could directly measure the escaping airstream; thirdly, we could ask the speaker what he thinks that he has articulated; and finally, we could ask a group of persons—who would here represent the speech community—whether aspiration was heard when the tape was played back.

It is beyond doubt that only the last method is of linguistic importance. The first and second methods are excluded, in lieu of all that has been said earlier. Moreover, measurements made on the oscillogram will never yield the binary decision; "present—not present," as is necessary for transcription purposes. All shades of curves resembling the curves typical for aspiration are found after $p\ t\ k$. The question as to whether and when they are perceived and evaluated as aspirations cannot be answered by the measurement itself. It is only on the basis of the decisions in terms of "observable—not observable," made during the listening, that the variation within both classes can be ascertained by means of numerous measurements.

The third case is excluded because the speech community is indifferent to what a person thinks he has said, if his words have not been heard. We can all be mistaken and can, in any kind of self-observation, be subject to autosuggestion.[7] Therefore we cannot base a linguistic tradition on the subjective conception of our own pronunciation.

Now the transcriptions are compared, the phonetic variants collected, and the cases in which no decision as to the presence or absence of aspiration could be reached are set apart in a third class of uncertain or doubtful cases. They are not included, for the purpose of measuring, in one of the first two groups. By investigating the phonotactic context of these instances of uncertain aspirations, we can then examine whether the context and the variability of the sound production have some common linguistic factor.

We will give an example of this procedure: In a New High German spoken text containing 20,000 sound symbols, 151 transcription variants were examined as to aspiration. A comparison of two such texts from the same speaker, recorded on phonograph records at an interval of there months, showed 136 differences in pronunciation relating to aspiration. Of these 136 cases, 35 occurred as transcription variants in the first text, 11 as uncertain aspirations in the second, and nine were uncertain and doubtful in the first text as well as in the second. This means that in 99 cases the same speaker certainly pronounced the same sounds with aspiration at the one time and unquestionably without aspiration at the other. Then it was necessary to examine in how many cases all three listeners certainly heard or did not hear the aspiration in both texts. In the material under observation, unequivocal aspiration in both phonetic texts was found in a total of 1127 cases and unequivocal absence in 1189 cases. Only this collection of data gives an overall view of the pronunciation of the speaker in question and characterizes it linguistically. This brief discussion may have been sufficient to indicate that such transcriptions provide a kind of cross-section of eventual changes in the usual pronunciation and—to use a botanical analogy—that this cross-section reveals a meristem or growing point. In other words, it allows us to distinguish types of pronunciation that are fluctuating from types with little variation.

These phonetic texts are now multiplied and utilized as a basis for the transcription of word stress, sentence stress, pitch contour (in sentences, words, syllables, and individual speech sounds), quantity, and speech pauses. In another work,[8] we set forth the method by which these factors can be transcribed and how the transcription results can be noted down in one of the multiplied phonetic texts by means of a system of diacritical signs, which are adapted to the language in question. In later publications, we have, in addition, shown how these phonetic texts are associated with different kinds of curves, or rather how the different kinds of curves are associated with phonetic texts. These curves can be oscillograms, logarithmic or linear intensity curves, registrations of fundamental frequency (so-called pitch curves), or sonagrams. They can also be physiological curves obtained by directly recording the movements of the speech organs or by quantitative interpretations of sound films and X-ray sound films.

We obtain a "phonometric text-list" by associating and adding to the phonetic text the data derived by the various measurements; for practical reasons it is written vertically in this case.[9]

We will now explain the abbreviations and the meaning of the figures in this text-list:[10] Above the table are three texts:

The orthographic text, which serves merely for rapid orientation.

The phonemic-allophonic text.

The phonetic text.

To the left of the table the phonemic-allophonic text and the phonetic text are repeated vertically. As has already been mentioned,[11] the phonetic text has been derived from the comparison of three transcriptions. Those passages which differed in one of the transcriptions have been noted on the "preliminary" list of variants; they have been reexamined twice by the transcribers; as a result, nine judgments are available for

1. Der Blick allein , der Römerart   u
2. deɐ blɪk ʔalaɪn , dee rømeɐʔart ʊ
3. deɐ blɪkh ʔalaɪn , dee RømeɐʔaRth ʊ

2.	3.	Q₁	Q₂	q₁	q₂	P	p	M	m₁	m₂	m₃	m₄	A₁	A₂	K/R	α₁/a₁	α₂/a₂	α₃/a₃
d	d																	
e	e			10	10			\\\	+2,1°	1,24	77,2	0,4			2	14	87,5	8,75
ɐ	ɐ																	
b	b			4											+11	1,14	0,93	0,85
l	l			11	9 (2)													
ɪ	ɪ			11	10(1)			///	+6,6°	1,14	86,7	1,0	/	///	9	16	81,5	7,41
k	k			6														
	h			6											−12	0,89	0,70	0,64
ʔ	ʔ																	
a	a			10	10			\\\	−1,0°	0,54	81,9	0,2			1	18	116,5	11,65
l	l			7	7										+7	1,06	2,18	1,11
ai	a/i			21	21			/ʌ⌐	+12,4°	1,06	81,4	4,4	/	\/\	6	19	254,0	12,95
n	n			11	9 (2)	)|||	36											
d	d																	
e	e			13	12(1)			Λ\Λ	−6,2°	0,66	75,4	1,2			2	14	131,0	10,08
ɐ	ɐ														+11	1,21	1,23	1,15
r	R			7	(1) 6													
ø	ø			14	14			///	+23,5°	1,00	83,5	5,7	/	///.	9	17	161,5	11,54
m	m			8	8										−12	0,90	1,19	1,19
e																		
	ɐ			14	7(7)			\\\	−31,3°	1,23	83,1	3,6			1	19	136,0	9,71
ɐ															+11	1,05	1,12	1,31
ʔ	ʔ																	
a	a			12	12			−\\	+31,0°	1,36	80,0	6,6	\	//\	8	20	152,5	12,71
r	R			5	4(1)										−11	3,33	7,44	3,10
t	t			6														
	h			1														
ʊ	ʊ			5	(1)4			\\\	+47,9°	1,30	74,2	3,3	/		1	6	20,5	4,40

them, while the remaining passages are based on three identical transcriptions. The majority vote has been incorporated into the phonetic text. The "variants" have been published in detail in the *Lesebuch neuhochdeutscher Texte*.[12]

Q_1—The first column in the list contains diacritical symbols representing the phonemic distinction of quantity: the opposition of long and short.

Q_2—The next column contains the transcription symbols for the *audible realization* of this opposition: its phonetic realization. The task of the listener was to examine whether the phonemic quantity was pronounced or is heard as particularly long or short. For this purpose, six diacritical marks are available to the transcriber, three to indicate shortness and three to indicate length:

1. ᷇: particularly short, 2. ᷇: short, 3. ᴗ: almost long;
1. ≡: particularly long, 2. =: long, 3. _: almost short.

q_1—The figures in this column give the *physical duration* of each sound that could be segmented on the kymographic intensity curve, expressed in ϕ (in 1/100 second). Sometimes the duration could be given only for two sounds taken together when a natural division could not be made. This was especially frequent in the case of diphthongs and the ending -*er*. The segmentation has always been established by associating the phonetic text with the kymogram on the one hand and with the embedding intensity curve[13] on the other. Normally, the segmentations obtained with the kymogram and the neurogram coincided completely. In the cases wherein the beginning or the ending of a sound could not be determined by means of the kymogram, the entries giving the physical duration have been left out (for example, [d]).

q_2—This column indicates the *duration of the periodic* segments of vowels and some voiced consonants *as established on the kymogram*. These figures are not too reliable, because of the resonance of the writing level of the kymograph that was used in 1936. However, they do show on which basis the pitch calculations of the following columns have been made. The kymographic method was the best method available at the time and was even suitable for voluminous pitch measurements; the equipment satisfied even high demands of precision. For these reasons, the pitch values have been included in the text-list. The figures given in brackets indicate the nonperiodic sections on the kymogram.

P—In this column the speech pauses or reading pauses are given, as they were obtained in the auditory transcription process. The hearers have at their disposal two diacritical signs:╱ short pause,╪ long pause. Each pause has to be assigned to one of these two classes.

p—The figures which are given beside the diacritical symbols indicating the pauses refer to the *measured duration of the pause*, which is again given in ϕ (in 1/100 sec). It happens sometimes that short pauses are measured which have not been noticed by the listeners while sometimes the opposite is observed—one of the listeners believes he hears a short pause for which the kymogram and the neurogram reveals no objective basis.

M—This column gives the transcription results with regard to the speech melody. The listeners were asked to characterize the pitch movement, not for the whole syllable, but, where feasible, only for the individual sound, by means of diacritical symbols. These signs were to be added to the phonetic text which was prepared beforehand. Numerous transcription exercises have shown that it is sometimes possible to ascertain in a syllable the pitch of the vowel and the following (more rarely the preceding) sound independently, especially if this is a liquid or a nasal. For the notation of these melodic contours by means of diacritical symbols, the following *melodic classes* were used by the transcribers: 1. falling \ , 2. rising / , 3. high-level ⌐ , 4. low-level ⌐ , 5. mid-level and further the following *combinatory classes*: 6. falling-low _ , 7. rising-high /⁻ , 8. high-falling ⁻\ , 9. low-rising _/, 10. rising-falling /\ and 11. falling-rising \/ . These are, if abstraction is made of combinations with the mid-level tone, all the possible combinations: five simple and six combined melodic classes.

m_1—This column shows the degree of rising or falling of the pitch as it has been calculated from the measurement of the periodic part of the kymogram by means of the trend method: This method makes it possible to calculate the main trend of a pitch curve on the basis of the numerical values underlying the graph. The result of these calculations of the trend of first degree is a line segment with a certain inclination as measured in degrees. Here this rise or fall is called the *pitch angle*. Thus the principal curve can be given for each sound as a line of which the beginning and end points are given in terms of time and height. (This is illustrated by Figure 7.)

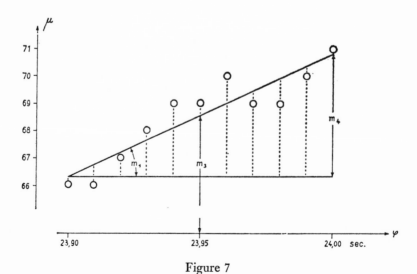

Figure 7

The horizontal line represents the duration—that is, the position in time—of the examined sound in the text; the vertical line represents the pitch in quarter tones (μ). The small circles indicate the pitch contour of the sound o (occurring in the phrase: "den Kern grosser Dinge" [14]). The hypotenuse of the rectangular triangle indicates the gradient of this pitch contour (trend of the first degree), the horizontal side of the triangle represents the duration of the sound q ($= q_1$) $= 10 \phi$, and the vertical side of the triangle represents the height (m_4) to which the sound rises.

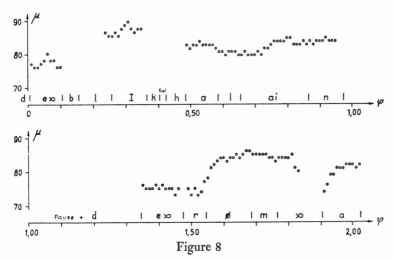

Figure 8

The beginning and end points themselves interest us neither in regard to their position in time nor in regard to the absolute pitch, which we always indicate in quarter tones (μ) above the lowest auditory threshold of 16 cps.

The first column of the melodic measurements indicates the *pitch angle* (m_1). This value is given by the angle at which the hypotenuse of the above figure rises above the horizontal axis ($+m_1$) or descends beneath it ($-m_1$). Theoretically, the values of m_1 can range from almost $+90°$ to almost $-90°$. The variation has been examined in my essay on pitch that was published in 1935.[15]

Figure 8 gives the measurements of μ for the present text-list.[16]

Figure 9 shows the graphical representation of the values of m_1 that have been calculated from the values of μ. In this way, the pitch contour of the whole text (including the syllabic, word, and sentence melody) can be deduced from the values of m_1 and represented graphically. The scientific value of such representations, however, is usually

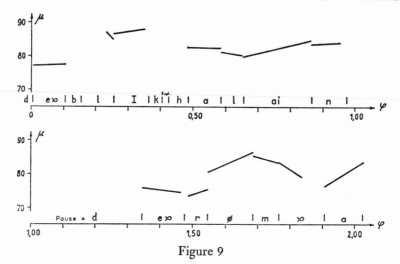

Figure 9

overestimated, as only statistical investigations can here yield results whose reliability can be tested.

m_2—This column indicates the mean deviation of the pitch melody: the mean square deviation of the pitch curve from the first degree trend. The values of m_2 have been given in units of quarter tones.[17]

m_3—Here the *average pitch of the vowel* in question has been given, again expressed in quarter tones. The figures given in this column show how many quarter tones the middle of the trend line lies above the lowest audible tone of 16 cps.

m_4—This column gives the *height of the pitch angle*: the vertical distance between the beginning and end point of the trend, again measured in quarter tones. The value of m_4 represents a measure for the rise or fall of the pitch within a sound. The value of m_4 is a function of m_1 and the duration of the sound.

A_1—This column shows the *word-stress*, so far as it is phonemically relevant. However, this remains a problem awaiting solution.

A_2—This column contains the transcriptions of three listeners concerning word-stress, expressed in a system of three degrees of stress: $\smile$ for unstressed syllables, $\diagdown$ for weakly stressed syllables, and $\diagup$ for strongly stressed syllables.

K/R—The figures in this column indicate the *stress class*[18] or *combination class*; that is, the class that results from the combination of the transcriptions produced by the three listeners. The underlined numbers between these figures give the so-called "relation class." This is the class in which the possible relations between the combination classes can be arranged from a linguistic point of view.

a_1, a_2, a_3 / a_1, a_2, a_3—In the previous column, the stress classes and the combination classes were given together with the relation classes, which expressed the linguistic relationship between the combination classes. In the same way, the last columns give the immediate values of the curve of average intensity. These values, corresponding to the combination classes, are denoted by the letter a. The values that correspond to the relation classes are denoted by the letter a; they give the values calculated on the basis of the a-values.

The measurements of stress have been based on a curve of the average intensity which was obtained by means of the neurograph of J. F. Tönnies.[19] This instrument registered the intensity in intervals of 1 mm = 1/100 sec, as the paper speed was 100 mm/sec.
Figure 10 indicates the shape of this curve for the recording that was also used for figures 8 and 9. The calculations of stress are based only on those parts of the curve that correspond to the vowels. These parts are indicated on the diagram by hatching.

By a_1 we designate the values of the so-called *peak speech power*. By a_2 we designate the *total speech power*; and by a_3 the *average speech power*.[20]
Figure 11 shows the average speech power for syllables of the sentence. The values a_1, a_2, a_3—which are given between the a-values—are the quotients of a.

Further statistical and arithmetic procedures which guarantee an appropriate correlation between physical intensity and the linguistic

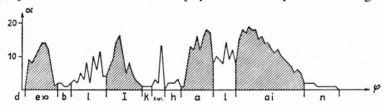

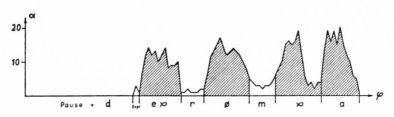

Figure 10

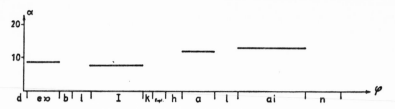

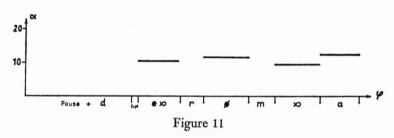

Figure 11

transcription of loudness or stress have been explained elsewhere,[21] together with the linguistic and mathematical rationale underlying our methods.

2. VARIATION OF PHONOMETRIC FEATURES

The preceding epistemological discussions have laid the foundation for phonometric statistics. Its principles will be described below.[1]

Charlier distinguishes between *homograde* and *heterograde* statistics.[2] Homograde statistics deals with the counting of a feature and the subsequent statistical analysis of the resulting numerical material. In heterograde statistics, however, the material has to be first measured and then arranged according to an artificially constructed classification. Only after this has been done can the number of occurrences of the classes be subjected to statistical analysis.

When a linguist says, for instance, "the /a:/ in German 'habe' is long, the /a/ in 'hatte' is short," both words are for him representatives of linguistic norms by which the language community abides—at least at a certain time. In phonometrics, however, when a tape recording and its oscillogram are examined we are dealing, not with linguistic norms, but with audible and visible phenomena that, as such, cannot be said to be valid in a language community or to have a historical development.[3] On the other hand, the judgment of the linguist rests on a number of observations characterized by the fact that they do not seek to ascertain the objective duration of sounds: the aim of these observations is to group the sounds heard into as many classes as seems linguistically warranted by the number of distinctive features.[4]

In both cases, the aim is to achieve a scientific understanding of a number of phenomena. Statistics provides a means to solve such problems and distinguishes in the same manner as was done above between the two procedures.[5] The phonetic question that is raised is, in one case, "Is it this sound or that one?" In other words, when writing a transcription, the listener decides into which of two or more determined classes a speech sound would be placed. In the other case, the question is asked, "How long is this particular sound, which has already been allocated to one of the given classes?"

In the first case, the problem is quantified by the simple counting of sounds. In the second only the measurement of individual sounds induces a quantification. In the first case, simply the presence or absence of one or more features has been established and counted within the linguistic classes. In Chapter III of this book, it was shown that such quantitative treatment[6] of speech sounds is justified. In the second case, the quantity of a variable feature has to be determined by a numerical value. The general necessity and justification of subjecting speech curves to such measurements has also been shown. This applies to the measurements of the movements of the speech organs as well as to the changes in the density of the air, which correspond to the act of speaking. In what follows below, our aim is to outline the general ideas on which such measurements and their statistical treatment are based.

To measure is to determine an empirical object by means of a geometrical norm on the appropriate scale.[7] It is therefore only possible to speak of measuring and of standards when there is a double relation between the variability of the object and a content determination satisfying the conditions of the law;[8] that is, when there is a relationship between the variability of a phonetic speech sound and its phonetic norm. These conditions must be fulfilled by all acoustic and physiological methods of measuring speech sounds and their results. This is true irrespective of whether we are dealing with problems of stress or quantity or with the investigation of the movements of the speech organs.

Another requirement for the statistical analysis of the results of the measurements is that the linguistic material has to be arranged in a suitable order. Objects that are arranged into one statistical class must always have a certain similarity. "The degree of similarity must be appropriate to the problem under examination. An insufficient degree of similarity can deprive the result of any scientific value whatsoever." [9]

Therefore, it is not *through* a statistical analysis, but *before* it, that the similar cases must be brought together. This is done in our case by establishing phonetic texts and determining the pitch, stress, and quantity classes, etc. Only listeners who know the language under investigation and who are therefore capable of discerning the linguistically relevant classes are entrusted with this task. Then the variations of variable features within these classes are determined by measurements and analyzed statistically.

The measuring procedures have been explained in previous essays[10] and are further elaborated in subsequent publications in relation to new problems. We will here merely elucidate the methods of heterograde statistics.

The statement, "Extreme cases are rare, while average cases are frequent," not only sounds plausible, it often even turns out to be true. Everybody has a clear idea of this state of affairs on hearing a statement like: "One seldom meets a dwarf or a giant; but people of normal or medium size are met very frequently." A critical examination of this statement is nevertheless necessary because the notions "extreme," "medium cases," and "frequently and rarely" are not defined with sufficient precision and are therefore ambiguous. This ambiguity can, however, be largely eliminated. Definitions of "extreme," "rare," "medium," and "frequent," adequate for statistical purposes, can be given. A numerical algorithm rests on an elementary geometrical arrangement of integers which are derived from one another by the simple arithmetical process of addition. In an equilateral triangle, the number 1 is placed in each angle. The second step consists in extending this basic triangle by adding as many lines as one wishes, parallel to the base and equidistant from each other. The number 1 is placed again in each of the angles thus formed:

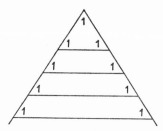

For the preliminary study of this useful algorithm, it is expedient to number the horizontal lines in such a manner that the top of the original triangle is included as line number zero, while its base is counted as the first line.

The third and last step is to fill in the triangular pattern systematically with integers which result from simple addition of two adjoining numbers that were obtained earlier by the same procedure. While the lines zero and one remain unaltered, the second line is completed by placing in the middle the figure which represents the sum of the two numbers directly above it to the right and the left. Hence, the sum of 1 and 1 is 2, which completes the second line. The third line is completed by placing two additional numbers which are again the sum of the figures directly above them to the right and the left. That is, $1 + 2 = 3$ and $2 + 1 = 3$. In this manner the algorithm continues from line to line and can be extended at will. The result of this procedure is as follows:

Pascal's Triangle

row		total of rows
0	1	0
1	1 1	2
2	1 2 1	4
3	1 3 3 1	8
4	1 4 6 4 1	16
5	1 5 10 10 5 1	32
6	1 6 15 20 15 6 1	64

That this numerical pattern is more than mere play becomes clear from the fact that in each line the sum of the numbers corresponds to the powers of two. The totals in each line—1, 2, 4, 8, 16, 32, 64—are precisely those powers of 2 that are indicated by the line numbers given on the left side of the triangle. Thus:

$$
\begin{aligned}
1 &= 2^0 \\
2 &= 2^1 \\
4 &= 2^2 \\
8 &= 2^3 \\
16 &= 2^4 \\
32 &= 2^5
\end{aligned}
$$

etc.

Another very remarkable fact is that the figures of Pascal's triangle correspond directly with the development of the powers of a sum of two terms, a so-called binomial. A comparison of these numbers with the well-known algebraic formulae:

$$
\begin{aligned}
(a+b)^0 &= 1 \\
(a+b)^1 &= 1 \times a + 1 \times b \\
(a+b)^2 &= 1 \times a^2 + 2 \times ab + 1 \times b^2 \\
(a+b)^3 &= 1 \times a^3 + 3 \times a^2b + 3 \times ab^2 + 1 \times b^3
\end{aligned}
$$

shows that the coefficients in these formulae correspond row by row with the numbers of Pascal's triangle. The numerical coefficients on the righthand side of the formulae just quoted are called "binomial numbers." The numerical pattern of Pascal's triangle has great heuristic value for algebra, geometry, and the study of permutations and probability.

A final remark has to be added on the pattern of binomial numbers. If in the binomial a + b both terms are made equal, in other words if a = b and, more specifically, if a = b = 1, then the numbers of our original Pascal's triangle must also result. As the binomial formula is valid for all values of a and b, there is nothing to prevent us from selecting, for instance, the values a = 0.9 and b = 0.1 (with the result that a + b = 1). The result is a numerical pattern which does not exhibit the beautiful symmetry that we have encountered so far but which is nevertheless built up according to mathematical "laws." Instead of the regular, symmetrical distribution, we encounter here a "skewed" pattern. Such "skewed distributions" have been encountered in phonometrics from the very beginning and have been taken into consideration. This has, for instance, been the case with the distribution of short vowels where a symmetrical distribution was already precluded because of their approximation to zero. For the distribution of "rare occurrences"—here rare sounds—we have introduced Bortkiewicz' "law of small numbers" into phonometrics.[11]

The following provides a further example: C. F. Gauss (1777–1855) introduced into the theory of errors—that is, into the theory concerned with the errors that are part of every measurement—the notion that the frequencies and magnitudes of these errors are subject to definite laws.[12] These laws are based on the following assumptions which accord with experience. The number of positive and negative errors are of equal frequency. Small errors are more frequent than large errors.

If, for instance, three causes give rise to the following "elementary errors" with the same ease:

        1st cause: −2, −1, 0
        2nd cause: −1, 0, 1
        3rd cause: −1, 0, 1, 2, 3

then their combination can result in the following combination of errors:

    (A) −4, −3, −2, −1, 0, 1, 2, 3, 4

The frequency of these ultimately resulting errors will be:

    (B) 1, 3, 6, 8, 9, 8, 6, 3, 1.

The graphical representation of both series of figures (A) and (B) suggest the characteristic bell-shaped Gaussian curve of errors; this is on the assumption that the size of the "elementary errors" is infinitely small while their frequency approaches large numbers.

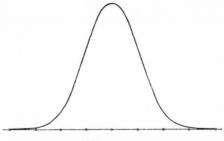

Figure 12

This theory was first introduced by Gauss into the technique of physical and astronomical measurements and in 1870 by Quetelet[13] into anthropometrics, a fact that has given rise to the development of biometrics—which in its turn has become, conjointly with the science of heredity, one of the most important fields of comparative biology. It is indeed a fact that

organisms, however closely related, are never totally identical. Anyone wishing to study what form these variations take needs to first examine each feature individually. The model for such investigations was provided long ago by Quetelet, who, among other things, has measured body height and many other body dimensions. Quetelet gives the heights of 26,000 soldiers, arranged in classes with intervals of one inch. In order to avoid figures with many digits, we have reduced the data to a sample of 1,000 men and tabulated the results below. The first row gives the measurements of height while the second row gives the number of soldiers in thousands.

Heights:	60″	61″	62″	63″	64″	65″	66″	67″	68″	69″	70″	71″	72″	73″	74″	75″ & above
Number:	2	2	20	48	75	117	137	157	140	121	80	57	26	13	5	3

It is clear that the individual heights are symmetrically grouped along both sides of the average height; the number of cases decreases gradually in both directions. After a closer examination of this and many other numerical series of a similar sort, Quetelet found that the distribution in the classes of such a table corresponds very well with the so-called binomial formula.

This so-called law of Quetelet, which states that the distribution of individuals approaches the binomial formula, has been partly confirmed by later investigators. It has thereby, however, become gradually apparent that the distributions do not always exhibit a pattern which is as simple and regular as Quetelet assumed.[14]

It could be assumed that what was valid for organisms would also be valid for the sounds produced by the speech organs. This claim has often been made since it was first put forward by Johann Conrad Amman more than 250 years ago; but it was first substantiated by the measurements that have been made by us over a period of thirty years. It is our opinion that the variation phenomena of spoken sounds and of their prosodic structures also provide a means to verify the validity of phonological oppositions on the level of speech.

Notes

ON BIBLIOGRAPHICAL QUOTATION

For the first citation of or reference to a work, the note gives the following information: author, title, year of publication of the edition used (italicized), and if necessary the year of publication of the first edition. This holds also for magazine articles and other dependent publications. Additional bibliographical information can be found in the Bibliography at the end of the book.

For subsequent citation of or reference to a work, the note gives only the author's name and the year of the edition used. If the author has more than one work in the same year, a letter follows the year [presumably *a* for the first work, *b* for the next, etc.]. Notes are numbered consecutively within each division of the book that is numbered in the Table of Contents (I, II.1, II.2, III.2, III.3, etc.).

II. *Observations On the History of Phonetics*

1. HISTORY OF THE TERM 'PHONETICS'

1. P.-J. Rousselot, *Principes de phonétique expérimentale.* 1st ed., 1897–1909 *(1924/25)*, vol. 1., p. 2, note 1: "La phonétique doit son nom à MM. Bréal et Baudry. Les introducteurs de cette science dans notre pays ont longtemps hésité entre les deux appellations phonétique et phonologie. Ils ont fini par rejeter la seconde, qui, avec notre transcription, peut signifier la science du meurtre (φόνος)" [φόνος: murder, homicide, blood-lust (also φονή); φωνή: voice, a) tone, sound, timbre, noise; b) speech, language, dialect, utterance, discourse].

2. F. Baudry: *Grammaire comparée des langues classiques* Pt. 1: *Phonétique (1868)*, footnote, p. 1, on the word "phonétique": "On dit aussi phonologie; c'est la Lautlehre des Allemands." A. Hatzfeld and A. Darmesteter: *Dictionnaire général de la langue française (1895)* designates "phonétique" as "néologisme admis par l'Académie française, 1878." The *Dictionnaire de l'Académie française*, 1878 ed., does not in addition list the word "phonologie". E. Gamillscheg: *Etymologisches Wörterbuch der französischen Sprache (1926–1929)* mentions "phonétique" only with the remark: "pertaining to the voice, 19th century, from Greek φωνητικός."

3. Presumably Rousselot is referring to M. Bréal: *Essai de Sémantique—Science des significations* (1st ed., *1897*), p. 9., "Je prie donc le lecteur de regarder

ce livre comme une simple Introduction à la science que j'ai proposé d'appeler la Sémantique"; and the note on this, "σημαντική τέχνη, la science des significations, du verbe σημαίνω (signifier), par opposition à la Phonétique, la science des sons."

4. Mentioned in the "Thesaurus graecae linguae," of whom the Stoics are probably the earliest and who after the Sophists may be regarded as the Greek founders of linguistics. In Latin, phonetics does not appear at all; in Greek only φωνητικός as an adjective, viz. φωνητικὰ ὄργανα; further φωνητικόν, neuter, also part of the soul (ψυχή) according to Plutarch among the Stoics, moral. K. E.-Borinski: *Grundzüge des Systems der artikulierten Phonetik zur Revision der Principien der Sprachwissenschaft (1891)*, p. 1, assumes that with φωνητική "according to Greek usage" τέχνη should be understood; only after that as medius terminus the ἐπιστήμη or θεωρία concerned with it. But Borinski too appears not to regard this development as old and as early as Greek. In all events, a noun φωνητική is unknown.

5. On the history of the hieroglyphic symbols cf. M. Rostovtzeff: *Geschichte der Alten Welt*, vol. 1 *(1941)*, p. 65: "At the earliest stage each single symbol—the Greeks speak of hieroglyphs—corresponds to a definite object: . . . ideographic or pictographic stage . . . syllabic stage. Finally, the syllabic characters very early became alphabetical characters or letters, each of which was attached to one of the twenty-four consonants of the Egyptian language. However, a purely phonetic style never prevailed; the Egyptians retained until comparatively late a complicated and clumsy combination of the three systems—ideographic, syllabic and alphabetic."

6. "But there are sufficient examples of the enigmatic class, and besides there is a fifth class of phonetic symbols which I have already urged can be referred to the enigmatic class; but since I find that it is neglected by most—although it is very useful in connection with what will have to be said below concerning the origin of the alphabet—I have considered that it must be treated separately." G. Zoega: *De origine et usu Obeliscorum (1797)*, p. 454. Georg Zoega (1755–1809) went to Rome in 1779—eleven years after Winckelmann's death—and there became the most highly regarded antiquarian. Cf. F. G. Welcker: *Zoegas Leben (1912–13)*; M. Russell: *Egypt XI (1853)*, p. 434: "To George Zoega also belongs the merit of employing (1797) the term phonetic"; further Renouf, P. Le Page: *Lectures on the Origin and Growth of Religion as Illustrated by the Religion of Ancient Egypt (1880)*, p. 13, note: "That the oval rings first contained royal names was first pointed out by the Danish scholar Zoega, who was also the first in modern times to assert that some hieroglyphic characters were phonetic." We will not attempt here a presentation of the changes in meaning of the numerous neologisms connected with phonetics in the 17th, 18th and 19th centuries. A good survey, at least of the English situation, appears in J. Murray (among others): *New English Dictionary on Historical Principles (1844–1933)*.

7. "The Rosetta stone shows us the use of this auxiliary writing system that we have called phonetics, i.e., expressing the sounds in the proper names of the kings Alexander, Ptolemeus, etc." J. F. Champollion, the Younger: *Lettre à M. Dacier . . . relative à l'alphabet des hiéroglyphes phonétiques employés par les Egyptiens etc. (1822)*, p. 4. It is obvious that Champollion knew the work by Zoega. He indeed does not quote from it, but he does cite other works by Zoega. Besides this, he relates his book from the beginning to that which preceded him, by J. D. Akerblad: "Lettre sur l'inscription égyptienne de Rosette" *(1802)*, which expressly refers to this work: "M. Zoega a entièrement épuisé cette matière dans son ouvrage De origine et usu Obeliscorum" (p. 57). Renouf

(1880): "The Swedish scholar Akerblad, already in the year 1802, drew up a phonetic alphabet of the demotic characters."

8. H. Salt: "Essay on Dr. Young's and Champollion's phonetic system of hieroglyphics" *(1825)*.

9. M. Russell: *View of Ancient and Modern Egypt with an Outline of its Natural History (1831)*, p. 184–91.

10. W. Kirby and W. Spence: *An Introduction to Entomology or, Elements of the Natural History of Insects*, vol. 4 *(1826)*, p. 331.

11. F. Bopp: *Vergleichende Grammatik*, 1st. ed. *(1833–52)*, p. 1428; further "a mere phonetic addition" (p. 1429), "a mere phonetic lengthening of the a and i" (p. 1430), or with reference to Greek ὑπέρ, Latin super in opposition to Old Indian upa-ri: "thus one is led to regard the Spiritus asper in Greek and the s in Latin in the prepositions concerned as a mere phonetic anacrusis." (p. 1473).

12. F. Bopp: *Vergleichende Grammatik*, 3rd. ed. *(1868–71)*, vol. 3, p. 452, note.

13. Bopp *(1868–71)*, p. 442, note 2.

14. W. von Humboldt: *Über die Verschiedenheit des menschlichen Sprachbaues und ihren Einfluss auf die geistige Entwicklung des Menschengeschlechts*, 1st. ed., 1836 (reprint *1935*), p. 89.

15. Humboldt *(1935)*, p. 90.

16. Humboldt *(1935)*, p. 102.

17. Humboldt *(1935)*, p. xi.

18. A.F. Pott: *Etymologische Späne (1856)*, p. 258.

19. F. Diez: *Grammatik der romanischen Sprachen*, Pt. I *(1836)*, p. 67. Cf. in this connection also G. Richert: *die Anfänge der romanischen Philologie und die deutsche Romantik (1914)*.

20. K. Burdach: *Die Wissenschaft von deutscher Sprache* (1934).

21. F.G. Klopstock: *Über die deutsche Rechtschreibung*, 1779 *(1830a)*; *Über Etymologie und Aussprache*, 1781 *(1830b)*; *Grundsätze und Zweck unsrer jetzigen Rechtschreibung*, 1782 *(1830c)*.

22. R. von Raumer: *Die Aspiration und die Lautverschiebung*, 1st ed., 1837 *(1863a.)*, p. 10.

23. "Rather this distribution of speech-sounds can only be discovered by the historical method; i.e., by the comparison of speech-sounds, by the observation of their change, and by their separation. They are, however, established by certain laws and participate in their nature phonetically, not dynamically, and thus are referable to their order." F. Delitzsch: *Jesurun, sive Prolegomenon in Concordantias Veteris Testamenti*, vol. 3 *(1838)*, p. 118.

24. K.M. Rapp: *Versuch einer Physiologie der Sprache*, vol. 1: *Die vergleichende Grammatik als Naturlehre. Erster oder physiologischer Theil (1836)*.

25. Rapp *(1836)*, p. 3.

26. H.E. Bindseil: *Abhandlungen zur allgemeinen vergleichenden Sprachlehre (1838)*, p. 5.

27. J. Müller: *Untersuchungen über die Physiologie des Menschen (1837a)*, *Handbuch der Physiologie des Menschen*, vol. 2, Pt. I. *(1837b)*, *Über die Compensation der physischen Kräfte am menschlichen Stimmorgan (1839)*.

28. M. Wocher: *Allgemeine Phonologie, oder natürliche Grammatik der menschlichen Sprache (1841)*, p. I; cf. also the article on Wocher by Lauchert in the *Allgemeine Deutsche Biographie*, vol. 43 *(1898)*.

29. A lengthy obituary appears in the journal *The Athenaeum*, London, March 17, 1888; a shorter notice listing his most important publications is to be

found in the *Dictionary of National Biography*, vol. 32 *(1892)*, p. 168–69. Latham's works are listed in the *General Catalogue of Printed Books* of the British Museum, vol. 131 *(1962)*, and in *Catalogue of Scientific Papers* (1867–1902), vol. 3 *(1869)*; cf. further also A.G. Kennedy: *A Bibliography of Writings on the English Language (1927)*.

30. R.G. Latham: *Facts and Observations Relating to the Science of Phonetics (1841a)*, p. 125ff.

31. R.G. Latham: *The English Language (1841b)*, p. 113; Latham was at this time Professor of English Language and Literature at University College, London. Cf. also Murray, amongst others *(1884–1933)*.

32. After 1900 it was called *Pitman's Phonetic Journal*, after 1905 *Pitman's Journal*, and since 1926 it has been called *Pitman's Journal of Commercial Education*. – Isaac Pitman (1813–1897) was among the champions of a "consistent" phonetic spelling in England, and in general one of the English reformers. In a "Report of meeting of phonographers in the Town Hall Manchester, July 14, 1868" *(Phonetic Journal*, August 1, *1868)*, there is a report of one of Pitman's lectures: "A history of phonography—How it came about"; in studying I. Walker's "A critical pronouncing dictionary, and exposition of the English language" (1824) Pitman arrived at "the first idea of the science of phonetics" around 1830.

33. Cf. especially W. Scherer: *Zur Regelung der deutschen Rechschreibung* 1869 *(1893a)*.

34. E. Brücke; *Grundzüge der Physiologie und Systematik der Sprachlaute für Linguisten und Taubstummenlehrer*, 1st ed., 1856 *(1876)*.

35. C.L. Merkel: *Anatomie und Physiologie des menschlichen Stimm- und Sprach-Organs (Anthropophonik)*, 1st ed., 1857 *(1863)*.

36. *Schmidt's Jahrbücher der gesamten Medicin*, vol. 95 *(1857)*.

37. *Zeitschrift für die österreichischen Gymnasien*, vol. 8 (1857).

38. Cf. also C.L. Merkel: *Die neueren Leistungen auf dem Gebiet der Laryngoskopie und Phonetik (1860)*.

39. On this cf. especially G. Panconcelli-Calzia: *Zur Geschichte des Kymographions (1935a)*.

40. cf. H. Breymann: *Die phonetische Literatur von 1876–1895 (1897)*. Very incomplete! The most important preliminary studies for a future history of experimental phonetics are G. Panconcelli-Calzia: *Zur Geschichte der phonoskopischen Vorrichtungen (1931)*, *Das Hören durch die Zähne (1934)*, *Der erste Kehlkopfspiegel: Babington's "Glottiskop"* (1829–1835) *(1935b)*. "One of the most successful researchers was Carl Ludwig (1816–1895), to whom we owe the invention of the Kymographion. The name Kymographion comes from A.W. Volkmann. With this instrument one can not only measure the blood-pressure, but also graphically represent its pulsatory variations." F.C. Müller: *Geschichte der organischen Naturwissenschaften in Neunzehnten Jahrhundert (1902)*, p. 122.

41. A. Ehrentreich: *Zur Quantität der Tonvokale im Modern-Englischen (1920)*, p. 6.

42. E. Sievers: *Grundzüge der Lautphysiologie*, 1st ed. *(1876)*; 2nd ed.: *Grundzüge der Phonetik (1881)*.

II. 2. ANCIENT PHONETICS

1. M. Müller: *Rig-Veda-Prātiśākhya, das älteste Lehrbuch der vedischen Phonetik (1869)*. Cf. also E. Bischoff: *Über die phonetische Systematik des San-*

skrit (1916) and T. Benfey: *Geschichte der Sprachwissenschaft und orientalischen Philologie in Deutschland (1869)*.

2. Brücke *(1876)*, p. 4.

3. E. Cassirer: *Philosophie der symbolischen Formen*, Pt. I: *Die Sprache*, 1st ed., 1923 *(1953)*.

4. L. Lersch: *Die Sprachphilosophie der Alten*, 3 parts *(1838, 1840, 1841)*.

5. R. Volkmann: *Die Rhetorik der Griechen und Römern (1885)*.

6. H. Steinthal: *Geschichte der Sprachwissenschaft bei den Griechen und Römern, mit besonderer Rücksicht auf die Logik*, 2 parts *(1890, 1891)*.

7. F. Stadelmann: *Erziehung und Unterricht bei den Griechen und Römern (1891)*.

8. A. Krumbacher: *Die Stimmbildung der Redner im Altertum bis auf die Zeit Quintilians (1921)*.

8a. A fuller treatment of the topic can be found in recent works and articles by W. S. Allen, D. Abercrombie, R. H. Robins, H. Koller, D. T. Langendoen, and others [translator's note].

9. Cassirer *(1953)*, p. 55 f.

10. J. Stenzel: *Philosophie der Sprache (1934)*. Cf. also *Über den Einfluss der griechischen Sprache auf die philosophische Begriffsbildung (1921)*.

11. E. Zwirner: *Das Gespräch. Beitrag zur Theorie der Sprache und der universitas litterarum (1951); Die Konsultation. Zur Theorie der psychophysischen Korrelationen und der "psychosomatischen Ganzheit" (1953)*.

12. Cassirer *(1953)*, p. 59.

13. Cassirer *(1953)*, p. 66 f.

14. Cf. Krumbacher *(1921)*, p. 63, note 1.

15. E. Frank: *Mathematik und Musik und der griechische Geist (1920/21)*.

16. *Plutarch de placit. Philos.*, from Lersch, Pt. 3 *(1841)*.

17. "Timaeus," 67b.

18. R. Westphal: *Aristoxenos von Tarent, Melik und Rhythmik des classischen Hellenentums*, 2 parts *(1883, 1893)*.

19. Dionysius of Halicarnassus: *De compositione verborum (1808)*, pp. 126 and 130. Cf. also R. Westphal: *Griechische Metrik (1887)* and Merkel (1863), p. 948, note.

20. Lersch, Pt. 3 *(1841)*.

21. E. Hoppe: *Geschichte der Physik (1926)*, p. 126.

22. Hoppe *(1926)*, p. 126 f.

23. *Aristotle probl. 41*, from Hoppe *(1926)*, p. 127.

II. 3. RENAISSANCE PHONETICS

1. Benfey *(1869)*.

2. W. Scherer: *Zur Geschichte der deutschen Sprache*, 1st ed. 1868 *(1878)*.

3. R. von Raumer: *Geschichte der germanischen Philologie, vorzugsweise in Deutschland (1870)*.

4. B. Delbrück: *Einleitung in das Sprachstudium (1893)*.

5. H. Paul: *Geschichte der germanischen Philologie (1891); Principles of the History of Language*: London, 1881, p. 39.

6. K. Burdach: *Vom Mittelalter zur Reformation*, 1st half *(1893); Deutsche Renaissance (1916); Reformation, Renaissance, Humanismus*, 1st ed. 1918 *(1963); Vorspiel*, vol. I, Pt. 2: *Reformation und Renaissance (1925)*.

7. Sievers *(1876)*.

8. Rousselot *(1924, 1925)*.

9. Cf. P. Hankamer: *Die Sprache, ihr Begriff und ihre Deutung im 16. und 17. Jahrhundert (1927)*.

10. Cassirer *(1953)*, p. 55.

11. Cassirer *(1953)*, p. 67.

12. ψυλικὸς δὲ ἄνθρωπος ου δέχεται τὰ τοῦ πνεύματος τοῦ θεοῦ. μωρία γὰρ αὐτ'ῷ ἐστιν, καὶ ου δύναται γνῶναι, ὅτι πνευματικῶς ἀνακρίνεται (I Cor. 2:14); cf. on this E. Zwirner: *Zum Begriff der Geschichte (1926)*, p. 49.

13. G. Ritter: *Studien zur Spätscholastik (1921, 1922)*.

14. Besides Galileo, it was the works of Kepler and the work of W. Gilbert "De magnete", first published in 1600 and frequently reprinted, that established the modern method of the natural science research. This book was used not only by Kepler, the only work of whom that had appeared before this was "Promodus," but also by Galileo, and was of influence on Descartes.

15. R. Hönigswald: *Die Philosophie von der Renaissance bis Kant (1923)*, p. 21 ff.

16. Cf. G. Woland: *Die moderne Grundlegungstheorie und der Galileische Naturbegriff (1964)* and E. Brüche: *Drei Kapitel Galilei (1964)*.

17. H. Arens: *Sprachwissenschaft (1955)*, p. 58.

18. Cf. on this E. Zwirner and K. Zwirner: *Lauthäufigkeit und Zufallsgesetz (1935a)*.

II. 4. PHONETICS OF THE SEVENTEENTH CENTURY

1. Cassirer *(1953)*, p. 68.

2. M. Lehnert: *Anfänge der wissenschaftlichen und praktischen Phonetik in England (1938)*.

3. Cassirer *(1953)*, p. 68.

4. 1724–1738.

5. "Sammlung und Abstammung Germanischer Wurzel-Wörter, nach der Reihe menschlicher Begriffe," Halle, 1776; "Versuch einer allgemeinen teutschen Idiotikensammlung," Berlin and Stettin, 1788; etc.

6. Cf. on this Paul *(1891)*, p. 53 f.

7. Biographical references can be found in Brücke *(1876)* p. 4; also in F. Neumann: *Die Taubstummen-Anstalt in Paris im Jahre 1822 (1827)*, p. 63.

8. J.P. Bonet: *Reduction de las letras y arte para enseñar a hablar los mudos* (1620) [quoted from Brücke *(1876)*].

9. J. Wallis: *De loquela, sive sonorum formatione, tractus grammatico-physicus (1653)*.

10. F.M. van Helmont: *Kurtzer Entwurff des Eigentlichen Natur-Alphabets der Heiligen Sprache*, 1667, reprint *(1916)*.

11. G.W. Leibniz: untitled essay in *Philosophische Schriften*, vol. 7 *(1890)*, p. 184. "There is an old saying that God has done everything by weight, measure and number. There are however things that cannot be weighed, e.g., things which have no force and power; there are also things which have no parts and to that extent cannot be measured. But there is nothing which cannot be counted. Therefore number is, as it were, a metaphysical figure and Arithmetic is a Static Element of the Universe by which the powers of things are explored."

12. Cf. T. Heuss: *Justus von Liebig* (1942).

13. G. C. Schelhamer: *De voce eiusque effectibus* (1677).

14. Narcissus, Lord Bishop of Ferns and Leighlin: "An introductory essay of the doctrine of sounds" *(1684)*, p. 472.

15. I. Newton: *Philosophiae naturalis principia mathematica (1687)*, p. 362.

16. Cf. further E. Hoppe: *Geschichte der Elektrizität (1884), Mathematik und Astronomie im klassischen Altertum (1911)*, and *Das antike Weltbild (1913)*.

17. T. Sheridan: *A course of Lectures on Elocution* (1762), *Lectures on the art of reading* (1775), *Über die Declamation oder den mündlichen Vortrag in der Prose and Poesie* (translation) *(1792)*.

18. C. G. Schocher: *Soll die Rede auf immer ein dunkler Gesang bleiben, und können ihre Arten, Gänge und Beugungen nicht anschaulich gemacht und nach der Art der Tonkunst gezeichnet werden? (1792)*.

19. C.E. Hamburg. The examples are printed in C.L. Merkel: *Physiologie der menschlichen Sprache (physiologische Laletik) (1866)*, p. 412 f.

20. C.H. Hänle: *Praktische, zum Theil auf Musik gegründete Anleitung zur Declamation und zum mündlichen Vortage (1815)*; here too the examples are in Merkel *(1866)*.

21. Merkel *(1866)*, p. 353.

22. C. F. Michaelis: *Die Kunst der rednerischen und theatralischen Declamation (1818)*.

23. E. Thürnagel: *Systematische Anleitung zur Declamation (1825), Theorie der Schauspielkunst (1836)*.

24. F.A.W. Diesterweg: *Beiträge zur Begründung der höhern Leselehre (1830)*. Further: L. Köhler: *Die Melodie der Sprache in ihrer Anwendung besonders auf das Lied und die Oper (1853)*.

II. 5. PHONETICS OF THE EIGHTEENTH CENTURY

1. J.C. Amman was born in Schaffhausen in 1669. Jespersen mistakenly calls him a Dutchman. Cf. O. Jespersen: *Fonetik (1897–99)*, p. 20.

2. Our italics.

3. Our italics.

4. *Phonetische Bibliothek*, No. 2 *(1917/18)*.

5. Viëtor's italics.

6. M. Dodart: *Memoire sur les causes de la voix de l'homme et de ses différents tons (1703)*.

7. J. C. Lischwitz: *Dissertatio de voce et loquela (1719)*.

8. A. Ferrein: *De la formation de la voix de l'homme (1744)*.

9. J.G. Runge: *Dissertatio inaug. de voce ejusque organis (1753)*.

10. C. de Brosches: *Traité de la formation méchanique des langues, et des principes physiques de l'etymologie (1765)*.

11. C.F. Hellwag: *De formatione loquelae (1781)*.

12. G.C.C. Storr: *De formatione loquelae (1781)*.

13. Cf. T. Meyer-Steineg and K. Sudhoff: *Geschichte der Medizin (1922)*, p. 338.

14. A. von Haller: *Anfangsgründe der Physiologie des menschlichen Körpers*, vol. 3 *(1766)*.

15. In the dissertation mentioned above *(1781)*.

16. O. Jespersen: *Lehrbuch der Phonetik (1926)*, p. 141.

17. 1742 – 1799.

18. G. C. Lichtenberg: *Vermischte Schriften*, vol. 8 *(1844)*, p. 171.

19. W. Ostwald: *Farbkunde (1923)*, p. 10 f., 13 f. Further; H.H. Weber: *Zur Frage der natürlichen Anordnung der Farben im Farbenkreis (1938)*, p. 86 f. (literature also given there); W. Ostwald: *Einführung in die Farbenlehre (1919), Die Harmonie der Formen (1922)*.

20. Meyer-Steineg und Sudhoff *(1922)*, p. 349 ff.

21. The "ventriculus laryngis (Morgagni)" or simply "Sinus Morgagni" is named after him. This laryngeal sinus, which allows the vibrations of the vocal-cords (ligamenta vocalia) free play, separates these from the upper "false" cords (ligamenta ventricularia), which are covered by a fold of mucous membrane and are not used in phonation.

22. After him is named the "cartilago cuneiformis (Wrisbergi)", a small rod-shaped laryngeal cartilage embedded in the tuberculum cuneiforme and which is embedded on each side in the plica aryepiglottica of the mucous membrane. Wrisberg's cartilage is – from the point of view of comparative anatomy – a descendant of the epiglottal cartilage. (Cf. E. Göppert: *Über die Herkunft des Wrisbergschen Knorpels*, 1894). According to J. Lossen (*Anat. Untersuchungen über die Cartilagines cuneiformes*, 1900), the cartilage is for half its length more or less rudimentary. It should be called more correctly Morgagni's cartilage.

23. Cf. R. Burckhardt: *Geschichte der Zoologie und ihrer wissenschaftlichen Probleme (1921)*; E. Rádl: *Geschichte der biologischen Theorien seit dem Ende des 17. Jahrhunderts*, Pt. 1 *(1905)*.

24. R. Hertwig: *Lehrbuch der Zoologie (1916)*, p. 9.

25. Ibid., p. 13

26. Discussing (in the Kantian sense of the concept of discussion) such relations with other disciplines after developing a scientific method, *post festum* so to speak, is different from making such analogies the guide to research.

27. Cf., e.g., J.P. Süssmilch: *Versuch eines Beweises, dass die erste Sprache ihren Ursprung nicht vom Menschen, sondern von Gott erhalten habe (1766)*.

28. A. Court de Gébelin: *Monde primitif (1775)*.

29. W. von Kempelen: *Mechanismus der menschlichen Sprache nebst der Beschreibung seiner sprechenden Maschine (1791)*, p. 31. Our interpolation.

30. Court de Gébelin *(1775)*, p. 74.

31. Burdach *(1934)*, p. 86, note 1.

32. E. Devrient: *Von der Entstehung des Stammbaums (1910)*, p. 2. Cf. also H.J. Bäumerich: *Über die Bedeutung der Genealogie in der römischen Literatur (1964)*.

33. Cf. Devrient *(1910)*; also O. Lorenz: *Lehrbuch der gesamten wissenschaftlichen Genealogie (1898)*, p. 87, note.

34. Cassirer *(1953)*, p. 70–102.

35. Cf. H. Swoboda: *Der Schachroboter. Herr von Kempelen und sein Automat (1965)*.

36. Burdach *(1934)*, p. 82 f.

37. The best exposition of this field is that of J. Schuster: *Die Anfänge der wissenschaftlichen Erforschung der Geschichte des Lebens durch Cuvier und Geoffroy Saint-Hilaire (1930/31)*. Schuster sees especially the relationship between Cuvier and Geoffroy in a different light.

38. Hertwig *(1916)*, p. 11 f.

39. Cf. K.E. von Baer: *Lebensgeschichte Cuvier's (1897)*; it contains an outstanding characterization of Cuvier and his period by an expert.

40. Cf. Burckhardt *(1921)*.

41. First ed. 4 vols. Paris 1812.

42. First ed. 4 vols. Paris 1817.

43. First ed. 5 vols. Paris 1801 ff.

44. First ed. 5 vols. Paris 1841–45.

45. Hertwig *(1916)*, p. 12

46. Cf. on this also Schuster *(1930/31)*.

47. Hertwig *(1916)*. p. 16–17.

48. G. Benn: *Goethe und die Naturwissenschaften, seine Kritik des Physikalismus* (1932).

49. Cf. R. Wagner: *Samuel Thomas Sömmerings Leben und Verkehr mit seinen Zeitgenossen (1844)*.

50. Besides Sömmering especially Camper, later Virchow (apart from Du Bois-Reymond); even Merck wrote of Goethe's "so-called discovery" (from J.H.F. Kohlbrugge: *Historisch-kritische Studien über Goethe als Naturforscher*, 1913); also on Camper, see Schuster *(1930/31)*.

51. On this cf. Kohlbrugge *(1913)*, p. 8 f.

52. J.H.F. Kohlbrugge: *Der Akademiestreit im Jahre 1831 (1919)*; but on this see also Schuster *(1930/31)*.

53. W. von Humboldt, in a long letter of 1795 in which he describes the Paris situation to him, reports enthusiastically of the Jardin des Plantes, "which is unique in Europe by virtue of its beautiful situation, the extent of the institution, the richness of the collections housed there, the scholarship and – one may add – the agreeableness of the scholars living there." "The most agreeable and most active man in this institute [he writes somewhat later] is Cuvier, who also knows German very well. He has done very interesting work on the physiology of cold-blooded animals, and intends to publish a detailed anatomia comparata. He lectures on this subject, and this course is supposed to be excellent"— *Goethes Briefwechsel mit Wilhelm und Alexander von Humboldt (1909)*, p. 52 f. Unfortunately, the catalogue of Goethe's library has still not been published. In his study there stood (and still stands) the first volume of the translation of Cuvier's *Thierreich*, published in 1831—that is, after the Academy dispute—by F.S. Voigt, the director of the botanical garden in Jena.

54. Kohlbrugge *(1931)*, as also J. Schuster: *Oken, Welt und Wesen, Werk und Wirkung (1929)*.

55. Cf. on this H. von Helmholtz: *Goethes naturwissenschaftliche Arbeiten (1896a)*; also *Goethes Vorahnungen kommender naturwissenschaftlicher Ideen (1896b)*.

56. On Goethe's theory of color and his relation to Newton cf. Ostwald (1923), p. 15.

57. From the point of view of the physiology and psychology of the senses, the achievement of the color-theory appears perhaps in a rather different light. On this cf. especially Jablonski, W.: *Zum Einfluss der Goetheschen Farbenlehre auf die physiologische und psychologische Optik der Folgezeit (1930)*.

58. F.H. Jacobi: *Auserlesener Briefwechsel*, vol. 2 *(1827)*, p. 368.

59. J.W. von Goethe: *F.H. Jacobis auserlesener Briefwechsel (1833)*, p. 293.

60. To what extent he did *not* mention, e.g., Vicq d'Azyr (of whom Huxley said: "He may be considered the founder of the modern science of anatomy"), has been shown by Kohlbrugge *(1913)*, p. 17 f.

61. Kohlbrugge *(1913)*, p. 23; further cf. also R. Friedenthal: *Goethe, Sein Leben und seine Zeit (1963)*, p. 354–70.

62. K. Vorländer: *Kant, Schiller, Goethe (1907)*; cf. also E. Zwirner: *Zum Begriff der Geschichte (1926)*, p. 48 f.

63. Kant: *Kritik der Urteilskraft*, 1st impression, 1790 *(1914)*, p. 497 f.; *Critique of Teleological Judgment*, tr. J. C. Meredith (Oxford: Clarendon, 1928), p. 78 ff.

64. Kant *(1914)*, p. 498.

65. Ibid., p. 498

66. Goethe met Blumenbach personally, on his visit to Göttingen in June 1801.

67. Cf. Vorländer *(1907)*.

68. Schiller's letters, vol. 3 *(1893)*, pp. 136, 186, 223.

69. J. Nadler: *Literaturgeschichte der deutschen Stämme und Landschaften,* vol. 3 *(1923)*, p. 272. Cf. on this also Goethe's correspondence with F.A. Wolf *(1868)*. Wolf paid a 14-day visit to Weimar at the end of 1803, the year in which Schelling and Loder left Jena. From October 22 to October 26, 1820, Wolf was in Jena with Goethe, "who lives there like a student." To this year belong Goethe's works *Zur Naturwissenschaft überhaupt, besonders zur Morphologie,* vol. 1, Pts. 2 and 3; in December of the previous year he had dictated his essay on the intermaxillary bone. Knowing how Goethe could draw those around him into the circle of his interests, one can gauge how strongly he will have interested F.A. Wolf—while Wolf was in Jena—in his anatomical ideas.

70. For details one should examine especially the passages referring to Humboldt in Burdach *(1934)* and W. Scherer's biography of Grimm *(1921)*; there too the influence of Humboldt and Grimm is described in detail; further, R. Haym: *Wilhelm von Humboldt, Lebensbild und Charakteristik (1856),* as also O. Harnack: *Die klassische Ästhetik der Deutschen (1892).*

71. After A. Riehl: *Galileo Galilei (1922),* p. 171 f.

72. Cf. p. 21 ff.

73. H. von Helmholtz: *Über das Sehen des Menschen (1896c),* p. 88.

74. One should read this in detail in the above-mentioned letters of Schiller to Körner of the beginning of 1793.

75. H. Junker: *Rede auf Wilhelm von Humboldt und die Sprachwissenschaft (1936).*

76. 1788. Quoted from W. Porzig: *Das Wunder der Sprache (1957),* p. 390.

77. *Vocabularia linguarum totius orbis comparativa, Augustissimae (Catharinae II) cura collectum (1786/89),* which contained 149 Asiatic and 51 European languages and dialects.

78. Quoted from Benfey *(1869)*, p. 268, note 1.

79. *Affinitas linguae hungaricae cum linguis fennicae originis grammatice demonstrata (1799).*

80. *Catálogo de las lenguas de las naciones conocidas y numeracion, division y clases de essas segun la diversidad de sus idiomas y dialectos (1800–1804).*

II. 6. PHONETICS OF THE NINETEENTH CENTURY

1. F. Schlegel: *Über die Sprache und Weisheit der Indier (1808),* p. x. Cf. on this also O. Jespersen: *Language, Its Nature, Development and Origin (1922).*

2. Schlegel *(1808)*, p. 3. Our italics.

3. E. Czuber: *Die statistischen Forschungsmethoden (1921),* p. 3. Our interpolation and italics.

4. "Von der grammatischen Structur."

5. Schlegel *(1808)*, p. 28.

6. *Codex Leicester 9v,* quoted from R. Weyl: *Die geologischen Studien Leonardo da Vincis (1950),* p. 273. Our interpolation.

7. 1638–1686.

8. *De solido intra solidum naturaliter contento dissertationis prodromus*

(1669), quoted from H. Hölder: *Geologie und Paläontologie (1960)*, p. 362. Our italics and interpolation.

9. Quoted from Hölder *(1960)*, p. 363. Our italics and interpolation.

10. "Discours sur les révolutions de la surface du globe, et sur les changements qu'elles ont produit dans le règne animal" (1st impression, 1812, as "Discours préliminaire..."), quoted from Hölder *(1960)*, p. 372.

11. Cf. Heuss (1942), pp. 15, 26.

12. Page 1 ff. Our italics in the quotation.

13. Page 17. On the preface to the *Vergleichende Grammatik*, cf. p. 83.

14. On this point cf. V. Thomsen: *Geschichte der Sprachwissenschaft (1927)*, p. 59 f.

15. Cf. W. von Humboldt: *Die sprachphilosophischen Werke (1883)*, p. 30 (the Körner quotation); further, the excellent epilogue on Humboldt by E. Wasmuth *(1935)*. W. Lammers: *Wilhelm von Humboldts Weg zur Sprachforschung 1783–1801 (1936)*, is thinking, in his work, less of linguistic research in the sense created by Grimm and Bopp but rather of Humboldt's philosophical occupation with language.

16. Cassirer *(1953)*, p. 97 ff.

17. On this point cf. E. Zwirner: *Die Bedeutung der Sprachstruktur für die Analyse des Sprechens. Problemgeschichtliche Erörterung (1965)*.

18. Benfey *(1869)*, p. 527, criticizes Humboldt's "not truly scientific" method; "it leads from periphery to periphery and is exposed to the danger ... of having exhausted its powers before it reaches the real nucleus of its task." In the final analysis, W. von Humboldt—as appears clearly from many passages in his letters—wanted less to promote research than to educate himself. This difference of attitude, or at least of emphasis, retains its weight, even though the gain for research, through the former, remains considerable.

19. It refers to the essay "Über das vergleichende Sprachstudium in Beziehung auf die verschiedenen Epochen der Sprachentwicklung," 1820 *(1963)*.

20. *Briefwechsel der Brüder Jacob und Wilhelm Grimm mit Karl Lachmann*, vol. 1 *(1927)*, p. 389.

21. Cf. above, pp. 53–54.

22. On Schlegel's organic conception cf. also Delbrück *(1893)*. The Romantic Movement obviously owes the application of the organism idea to all intellectual and natural being and life to Schelling's nature philosophy, although Schlegel had basically learnt nothing from Schelling's transcendental philosophy. That was also Dilthey's view (U. Joachimi: *Die Weltanschauung der deutschen Romantik*, 1905). Certainly this conception was in the air then; it was contained in the micro-macrocosm doctrine of the Greeks, in the monad-theory, in aesthetic evaluations of Herder (F. Koch: *Herder und die Mystik, 1927/28*, p. 11). and of the young Goethe, who gladly made this organism-concept the basis of these evaluations. And J.W. Ritter's discovery of the electrical nature of Galvanism ("Beiträge zur näheren Kenntnis des Galvanismus," 1800–1805) seemed, quite superfluously, to confirm the "organic-intellectual unity" of nature just at the right time, so to speak, experimentally and thus quite irrefutably! The general interest in these questions and the superficial approach to the problems associated with it, was probably much greater then than was wholesome for natural science.

23. Schlegel (1808), p. 51.

24. In the foreword to the *Critique of Pure Reason (1787)* he says: "This experiment of pure reason has much in common with that of the chemists. . . ."

(p. xx, note). The anonymous editor of J.G. Jacobi's *Sämmtliche Werke (1825)*, wrote: "The intellectually kindred men attracted each other irresistibly, just as certain chemical elements easily approach each other and merge" (vol. 1., p. 39). Even some time later chemistry had similar effects—even on physics: Robert Mayer, e.g., continually refers to Lavoisier's example between 1841 and 1871. In August 1841 he wrote to Baur: "The same situation as in the theory of matter (chemistry) applies also to the theory of forces (physics); both must be based on the same principles."

25. Cf. M. Scholtz: *Lehrbuch der pharmazeutischen Chemie*, vol. 1: *Anorganischer Teil (1910)*, p. 31 f.; further, F. Paneth: *Über die erkenntnistheoretische Stellung des chemischen Elementbegriffs (1931)*.

26. Brücke *(1876)*, p. 7

27. Cf. W. Scherer: Review of "*Grundzüge der Physiologie und Systematik der Sprachlaute*" by E. Brücke *(1893b)*.

28. Cf. Kempelen *(1791)*, p. 35: "and so on, back to an original language, but the records for this are lacking."

29. Kempelen *(1791)*, p. 49 f.

30. J. Grimm: Deutsche Grammatik, pt. 1, 3rd impression (1840), p. xv.

31. Merkel *(1866)*, p. iv.

32. 1st impression *(1836)*.

33. W. von Humboldt and Goethe use "Töne" (apparently only in the plural) instead of "Laute" in the sense of "Sprachlaute." Humboldt speaks, e.g., of "willkürlich verabredete Töne." P. Fischer *(Goethe-Wortschatz*, 1929) does not list "Laut" and "Sprachlaut," nor "Ton" in the sense of speech-sound. Cf. however, e.g., *Wanderjahre*, book 1, ch. 3: "Letters may be a fine thing, and yet are inadequate to express the sounds; we cannot dispense with sounds, and yet they are by a long way not sufficient to allow the real meaning to be heard; in the end we cling to letter and sound; and are no better off than if we dispensed with them entirely; what we communicate, what is handed over to us, is only the most common thing, which is not worth the trouble" (p.45). This chapter may give a further insight into Goethe's poetic self-sufficient way of investigating the world with his eyes, which does not know the principle of discursive research and consequently in the last resort denies the possibility of agreement: "If I now treated these very (geological) fissures and clefts as letters, attempted to decipher them, to form them into words . . . Nature has only One Script . . . 'And yet even here people will dispute one's readings.' 'For this very reason I speak with nobody about it' " (p. 46).

34. K. Burdach: *Einleitung zum "Briefwechsel der Brüder Jacob und Wilhelm Grimm mit Karl Lachmann" (1927)*, p. xxii.

35. Rapp *(1836)*, p. 171.

36. Cf. F. von Thiersch: *Gedächtnisrede auf Schmeller (1852)*; O. Föringer: *Lebensskizze Schmellers (1855)*; K. Hofmann: *Andreas Schmeller (1855)*; J. Nicklas: *J.A. Schmellers Leben und Wirken (1885)*. Further, cf. L. Rockinger in *Oberbayerisches Archiv*, vol. 43 *(1886)*; E. Schröder in *Allgemeine Deutsche Biographie*, vol. 31 *(1890)*; G. Könnecke in *Bilderatlas zur Geschichte der deutschen National-Literatur (1895)*; U. Koch in *Der Wächter*, 6 (1923).

37. Raumer *(1863a)*, p. 9 f. Cf. p. 21.

38. Süssmilch, J.P.: *Die Göttliche Ordnung in den Veränderungen des menschlichen Geschlechts, aus der Geburt, dem Tode und der Fortpflanzung desselben erwiesen (1741)*.

39. Gauss, K.F.: *Theoria combinationis observationum erroribus minimis obnoxiae (1823)*.

40. Raumer *(1863a)*, p. 88.

41. von Raumer, R.: *Offener Brief an den Herausgeber der Zeitschrift für die deutschen Mundarten*, 1857 *(1863b)*.

42. The daguerreotype was invented in 1838 by the French painter L.J.M. Daguerre and was used in Germany soon afterwards.

43. Page 364 ff. Our italics.

44. Page 366 f. Our italics.

45. Cf. on this also F. Trendelenburg: *Ohms akustiches Gesetz (1939)*.

46. F. Auerbach: *Tonkunst und Bildende Kunst vom Standpunkte des Naturforschers, Parallelen und Kontraste (1924)*, p. 78 ff.

47. A. Kalähne: *Grundzüge der mathematisch-physikalischen Akustik (1910/1913)*.

48. I. B. Crandall: *Theory of Vibrating Systems and Sound (1926)*.

49. H. Fletcher: *Speech and Hearing (1929)*.

50. F. Trendelenburg (ed.): *Akustik (1929)*.

51. Cf. H. Paul: *Prinzipien der Sprachgeschichte*, 1st impression, 1880 *(1920)*, p. 49 ff.

52. G. von der Gabelentz: *Die Sprachwissenschaft, ihre Aufgaben, Methoden und bisherigen Ergebnisse*, 1st impression, 1891 *(1901)*.

53. F. de Saussure: *Cours de linguistique générale*, 1st impression 1916. This posthumous work is based on Saussure's lectures from 1906 to 1911.

54. Page 27.

55. Page 29.

56. Page 9 f. Our italics.

57. Page 61.

58. Page 63.

59. Page 15. Our italics.

60. Page 83.

61. Page 89.

62. F. de Saussure: *Grundfragen der allgemeinen Sprachwissenschaft (1931)*, p. 119.

63. Saussure *(1931)*, S. 120.

64. J. Winteler: *Die Kerenzer Mundart des Kantons Glarus in ihren Grundzügen dargestellt (1876)*, p. 12. Our italics.

65. N. S. Trubetzkoy: *Grundzüge der Phonologie*, 1st impression, 1939 *(1958a)*, p. 8.

66. Cf. E. Zwirner: *Quantität, Lautdauerschätzung und Lautkurvenmessung (Theorie und Material) (1933a)*.

67. E. Zwirner: *Phonologie und Phonetik (1939)*, *Langue et parole en phonométrie (1938)*.

68. Cf. on this H. Schuchardt-Brevier, 1st impression, 1922 *(1928)*, p. 315 f.: "Romance Philology is, with regard to its limitation, a University subject, not an individual science...."

69. Characterized by G. T. Fechner a hundred years earlier as "external and internal psychophysics." Cf. *Elemente der Psychophysik (1859/60)*, ch. II: "Begriff und Aufgabe der Psychophysik."

70. Cf. e.g., A. Riehl: *Der philosophische Kritizismus*, vol. I *(1876)*, p. 375–575. Riehl began with Herbart's realistic foundation of the concepts of experience and cognition (*Realistische Grundzüge*, 1870), but then turned towards

the Critical movement which had its roots in Kant (*Über Begriff und Form der Philosophie*, 1872). For him as for Kant, physics was the model science. However, what is right for physics, is right for linguistics. And, just as the romantic philosophy of nature was overcome through Kant's theory of physical research before it began, the theory of linguistic research overcomes a language philosophy which knows neither its object nor its method. Cf. also R. Hönigswald: *Die Skepsis in Philosophie und Wissenschaft* (1914), and *Immanuel Kant (1924)*.

71. M. von Laue: *Erkenntnistheorie und Relativitätstheorie (1960)*, p. 62.

72. E. W. Scripture: Review of P. W. Bridgman, "The logic of physics" *(1930/31a)*.

73. C.F.P. Stutterheim: *Wijsbegeerte en Taalwetenschap. Problemen der Fonologie (1959)*.

74. Raumer *(1863a)*, p. 6.

75. Paul *(1920)*, pp. 54–63. There Paul speaks of "fluctuations", of conditioned "uniformity," "variability of pronunciation," and of "control of the feeling of motion by the sound-image" (p. 58).

76. E. Zwirner *(1933a)*. G. K. Zipf: *Phonometry, Phonology and Dynamic Philology and an Attempted Synthesis (1938)*.

77. L. A. Quetelet: *Anthropométrie, ou mesure des différentes facultés de l'homme (1870)*.

78. E. Zwirner *(1933a)*.

79. E. Zwirner, A. Maack and W. Bethge: *Vergleichende Untersuchungen über konstitutive Formen deutscher Mundarten (1956)*.

80. J. Grimm *(1840)*, p. 32.

III. *Methodological Foundations of Phonometrics*

1. Cf. D. Hilbert: *Über den Zahlbegriff (1900), Über die Grundlagen der Logik und der Arithmetik (1905)*. H. Rickert: *Das Eine, die Einheit und die Eins (1924)*.

2. Cf. G. Cantor: *Grundlage einer allgemeinen Mannigfaltigkeitslehre* (1883); *Beiträge zur Begründung der transfiniten Mengenlehre, I und II (1895/ 97)*. Further, A. Fraenkel: *Einleitung in die Mengenlehre (1928)*; F. Hausdorff: *Mengenlehre*, 2nd impression *(1927)*.

III. 1. TWO OBJECTIONS TO THE DISTINCTION AND DISTRIBUTION OF LINGUISTIC SEGMENTS

1. Cf. below, p. 107.

2. P. Menzerath and A. de Lacerda: *Koartikulation, Steuerung und Lautabgrenzung (1933)*.

3. 1st impression, 2 vols., 1897/1909.

4. 1st impression, 1880.

5. Paul *(1920)*, p. 51 f. Paul's italics.

6. What has been spoken, has been determined by listening to the recording. The fact that the curve under discussion corresponds to this and has been decided by a coordination procedure, whose presuppositions have still to be discussed.

7. Cf. below p. 105f.

8. V. Gottheiner and E. Zwirner: *Die Verwendung des Röntgenfilms für die Sprachforschung (1933)*. E. Zwirner: *Gestikulationskurven (1933b)*.

9. The logical basis for the "only" can be found in B. Riemann's *Habilitationsschrift* "Über die Hypothesen, welche der Geometrie zu Grunde liegen," 1854 *(1919)*.

III. 2. SOUND CLASS AND SOUND REALIZATION

1. W. von Wartburg: *Einführung in Problematik und Methode der Sprachwissenschaft* (1962), p. 196. The second sentence is the footnote to the first. Our interpolation.

2. F. Krüger: *Beziehungen der experimentellen Phonetik zur Psychologie (1907)*; O. Klemm: *Wahrnehmungsanalyse (1921)*; Schaefer, K.: *Psychologische Akustik (1922)*; E. R. Jaensch: *Untersuchungen über Grundfragen der Akustik und Tonpsychologie (1929)*; V. Engelhardt and E. Gehrke: *Vokalstudien (1930)*.

3. O. Weiss: *Stimmapparat des Menschen (1931)*. Cf. beyond this the biological viewpoints of A. Jellinek: *Die funktionelle Einordnung der Organismen in die Schallwelt (1932)*.

4. K. Stumpf: *Die Sprachlaute (1926)*; W. Sulze: *Die physikalische Analyse der Stimm- und Sprachlaute (1931)*.

5. Cf. on this – although deviating somewhat from our task through the tendency of the "style" – the "separation" of language from the speech act, of linguistics from the science of speech" by E. Otto in *Grundfragen der Linguistik (1934)*.

6. Cf. Gabelentz *(1901)*, p. v.

7. Cf. E. Staiger: *Die Zeit als Einbildungskraft des Dichters (1939)*, *Grundbegriffe der Poetik (1946)*, and *Die Kunst der Interpretation (1955)*; W. Kayser: *Das sprachliche Kunstwerk (1948)*.

8. Cf. E. R. Curtius: *Europäische Literatur und lateinisches Mittelalter (1948)*.

9. Cf. E. Oksaar: *Semantische Studien im Sinnbereich der Schnelligkeit (1958)*.

III. 3. THE SOUND UNIT (SOUND SEGMENT)

1. R. Hönigswald: *Grundfagen der Erkenntnistheorie (1931)*, p. 114.

2. Cf. Fletcher *(1929)*.

3. R. Hönigswald: *Die Grundlagen der Denkpsychologie (1925)*, p. 329 f. and Hönigswald *(1931)*, p. 85f.

4. Cf. in this connection Louis Hjelmslev's concept of "commutation."

5. Cf. Paul *(1920)*, ch. IV, ch. VII; Hönigswald *(1925)*; J. Stenzel: *Sinn, Bedeutung, Begriff, Definition (1925)*.

6. Cf. Hönigswald *(1925)*, p. 277.

7. Ibid., p. 252 f.

8. Cf. E. Zwirner: *Die Grundlagen der vergleichenden Sprachphysiologie und Sprachphysik*, Pt. 1 *(1934)*, pp. 155, 172.

9. Hönigswald *(1925)*, p. 81.

10. Ibid., p. 82.

11. Ibid., p. 83.

12. Ibid., p. 101.

13. Cf., e.g., Kant: *Was heisst: sich im Denken orientieren? (1922)*.

14. Cf. M. Isserlin: *Die pathologische Physiologie der Sprache (1929/31)*.

15. Cf. on this M. Löwe: *Schwellenuntersuchungen (1924)*.

16. V. Kraft: *Erkenntnislehre (1960)*, p. 50f. Kraft's italics.

17. Jespersen *(1926)*, p. 11f.

18. 1st impression, 1884.

19. p. 75.

20. *(1910)*, cf. particularly p. 183.

21. Viëtor *(1923)*, p. 77.

22. Scripture *(1930/31a)*.

23. H. Gutzmann: *Röntgenfilmaufnahmen des Sprechens (1930)*.

24. Scripture *(1930/31b)*.

25. Cf. E. Zwirner *(1934)*.

26. Cf. E. Zwirner *(1933b)*.

27. Cf. Weiss *(1931)*, p. 1294.

28. Cf. on this Viëtor *(1923)*, p. 44, as also Jespersen *(1926)*, p. 25. However, the question is not, whether we can speak without moving the lower jaw, but whether we usually do so in a particular language.

29. Cf. on this E. Cassirer: *Substanzbegriff und Funktionsbegriff* (1923).

30. Cf. M. von Rohr: *Das photographische Objektiv (1927)* and *Die Strahlenbegrenzung in ihrer Bedeutung für die Lichtbilder (1932)*.

31. Cf. W. Nagel: *Die Wirkungen des Lichtes auf die Netzhaut (1905)*.

32. Cf. D. Hilbert: *Grundlagen der Geometrie (1922)*, as also A. J. Dietrich: *Kants Begriff des Ganzen in seiner Raum-Zeit Lehre und das Verhältnis zu Leibniz (1916)*.

33. Cf. E. Zwirner *(1933b)*.

34. Cf. E. Zwirner *(1934)*.

35. Cf. E. Zwirner *(1933b)*.

36. H. Schole: *Experimentelle Untersuchungen an höchsten und an kürzesten Tönen (1934)*.

37. Cf. A. Gemelli and G. Pastori: *L'analisi elettroacustica del linguaggio. (1934)*.

IV. *Phonometric View of the Sound System*

IV. 1. SYSTEM OF DISTINCTIONS AND OPPOSITIONS

1. Gabelentz *(1901)*.

2. I. Fónagy: *Die Metaphern in der Phonetik (1963)*.

3. Jespersen *(1926)*, p. 109f.

4. N. S. Trubetzkoy: *Anleitung zu phonologischen Beschreibungen (1958b)*, p. 9 [trans.: *Introduction to Phonological Descriptions*, p. 7].

5. Trubetzkoy *(1958a)*, p. 30 ff.

6. *De vulgari eloquentia* (ca. 1305), quoted by Arens *(1955)*, p. 44.

IV. 2. SYSTEM OF SEGMENTS AND OF VARIANTS

1. Cf. Saussure *(1931)*, p. 44f.

2. Gabelentz *(1901)*, p. 5, cf. above, p. 123.

3. Hönigswald *(1925)*.

4. Cf. E. Zwirner and K. Zwirner *(1935a)*.

5. Trubetzkoy *(1958a)*.

6. Therefore, whether we write a strictly phonemic text with a suitable legend or an allophonic text amounts in practice to the same thing.

V. *Execution of Phonometric Procedures*

V. 1. COMPILATION OF TRANSCRIBED TEXTS AND
PHONOMETRIC TEXT LISTS

1. On this point see E. Zwirner: *Anleitung zu sprachwissenschaftlichen Tonbandaufnahmen (1964a)*.

2. E. Zwirner: *Die Beziehung der Phonemtheorie Trubetzkoys zur Phonetik (1964b)*, p. 14.

3. Cf. Trubetzkoy *(1958a)*, pp. 7f., 11 ff; further Trubetzkoy *(1958b)*, p. 5 [or *Introduction*, p. 1].

4. We speak here of "phonemic" in relation to "prosodemic" or "tonemic" but we do not reject out of hand the objections which H. Pilch has made against these terms in his *Phonemtheorie*, Vol. I. *(1967)*, p. xiii.

5. In the same way as manuscripts which are utilized for the production of a text are referred to by the capital letters A, B, C . . . , and the corresponding printed versions by a, b, c . . . , we refer to the transcriptions which are used for the production of a phonometric text by the numbers 1, 2, 3 . . . .

6. Cf. E. Zwirner and K. Zwirner *(1935a)*.

7. Cf. W. Wolff: *Selbstbeurteilung und Fremdbeurteilung im wissentlichen und unwissentlichen Versuch (1932)*.

8. Cf. E. Zwirner and H. Richter: *Gesprochene Sprache (1966)*.

9. Of such text-lists the following have been published: E. Zwirner and K. Zwirner: *Textliste neuhochdeutscher Vorlesesprache schlesischer Färbung (1936a)*; E. Zwirner: *Textliste märkischer Mundart (1936a)*, *Textliste schlesischer Mundart (1936b)*, and *Textliste neuhochdeutscher Mundart bayrischer Färbung (1937)*.

10. The gramophone record, on which this text-list is based, was made in October, 1933. Cf. E. Zwirner and K. Zwirner *(1936a)*, p. 41.

11. Page 139 ff.

12. E. Zwirner and K. Zwirner: *Lesebuch neuhochdeutscher Texte (1937a)*.

13. Cf. J. F. Tönnies: *Der Neurograph, ein Apparat zur Aufzeichnung bioelektrischer Vorgänge unter Ausschaltung der photographischen Kurvendarstellung (1932)*; J. A. Schaeder: *Zur Schallstärke- und Lautstärke-Registrierung für phonometrische Zwecke (1939)*.

14. Cf. E. Zwirner and K. Zwirner *(1936a)*, p. 32.

15. E. Zwirner and K. Zwirner: *Phonometrischer Beitrag zur Frage der neuhochdeutschen Lautmelodie (1935b)*.

16. Cf. E. Zwirner and K. Zwirner *(1936a)*.

17. E. Zwirner and K. Zwirner: *Phonometrischer Beitrag zur Frage des neuhochdeutschen Akzents (1936b)*.

18. E. Zwirner: *Phonetische Tonhöhenbezifferung (1933c)*.

19. Tönnies *(1932)*.

20. Fletcher *(1929)*, p. 64 ff; cf. also C. B. Miller: *Accent: Classes and Variations (1936)*.

21. E. Zwirner and K. Zwirner *(1936a)*.

V. 2. VARIATION OF PHONOMETRIC FEATURES

1. If phonometrics sees in statistics its specific method of research, then this is based on its epistemological understanding of its object. This connection was, however, often not understood as is evidenced by the fact that phonometrical

statistics was often identified with the method which had already long been used in pathology and which was applied in 1925 by R. Schilling to speech pathology (cf. O. Von Essen: *Allgemeine und angewandte Phonetik*, 3rd ed. 1962, p. 6–7). The same misunderstanding is responsible for the view which considers phonometrics as an "unsuccessful" attempt to measure and statistically calculate (phonological) norms (cf. E. Dieth: *Vademecum der Phonetik*, 1st edition, 1950, p. 19). This completely disregarded the fact that Schilling did not even tackle the problem of the linguistically based segmentation, let alone solve it. On the other hand, a claim has been attributed to phonometrics which it had rejected expressly 13 years before the appearance of the *Vademecum der Phonetik:* no amount of measurements or calculation tricks will allow us to "calculate phonemes" (cf. K. Zwirner: *Das Eindringen statistischer Forschungsmethoden in die Sprachvergleichung*, 1937). The way in which phonometrics is understood in Germany seems to us to be characteristic of the way in which phonetics in general is understood. Even today we find in textbooks on phonetics that the old claims of the school of experimental phonetics that it is scientific in the same manner as the natural sciences are retained while phonology and the structural approach to speech sounds are relegated to an altogether secondary place instead of constituting the point of departure of the discussion.

2. Cf. C.V.L. Charlier: *Vorlesungen über die Grundzüge der mathematischen Statistik (1920)*.

3. Cf. Paul *(1920)*, p. 28.

4. Cf. E. Zwirner *(1933a)*.

5. Cf. Czuber *(1921)*, p. 1.

6. Cf. E. Zwirner and K. Zwirner *(1935a)*.

7. Cf. R. Hönigswald: *Zur Frage: Nichteuklidische Geometrien und Raumbestimmung durch Messung (1915)*.

8. Cf. A. Kneser: *Mathematik und Natur (1918)*.

9. Czuber *(1921)*, p. 2.

10. E. Zwirner and K. Zwirner *(1935b)*, *Streuung sprachlicher Merkmale (1936c)*, *(1936b)*, *Phonometrischer Beitrag zur Frage der neuhochdeutschen Quantität (1937b)*.

11. L. von Bortkiewicz: *Das Gesetz der kleinen Zahlen (1898)*. Cf. E. Zwirner and K. Zwirner *(1935a)*.

12. Cf. K. R. Biermann: *Aus der Entstehung der Fachsprache der Wahrscheinlichkeitsrechnung (1965)*.

13. Quetelet *(1870)*.

14. W. Johannsen: *Elemente der exakten Erblichkeitslehre (1926)*, p. 8ff.

Bibliography

In the following list, the literature quoted in the text and notes of this volume has been compiled and completed bibliographically. Works which have been mentioned only as to their historical dates have been omitted. The date in italics refers to the edition used here. (See also page 156, "On Bibliographical Quotation.") As often as possible, the abridged title of journals is followed by the year or volume and then by the first and last page of the article in question. The name of the publishing house is given for independent works published after 1945.

Åkerblad, J.D.: Lettre sur l'inscription égyptienne de Rosette (Paris *1802*).

Amman, J.C.: Dissertatio de loquela etc. (1700). Reprint with German translation by G. Venzky (1747), (ed. W. Viëtor). Vox *27/28:* Separafum=Phonet. Bibl. No. 2 *(1917/18)*.

Arens, H.: Sprachwissenschaft. Der Gang ihrer Entwicklung von der Antike bis zur Gegenwart. Orbis Academicus, Vol. I/6 (Alber, Freiburg/ Munich *1955*).

Auerbach, F.: Tonkunst und Bildende Kunst vom Standpunkte des Naturforschers. Parallelen und Kontraste (Jena *1924*).

v. Baer, K.E.: Lebensgeschichte Cuvier's. Ed. L. Stieda. Arch. Anthrop. *24: 227–275 (1897)*.

Bäumerich, H.J.: Über die Bedeutung der Genealogie in der römischen Literatur. Thesis (Köln *1964*).

Baudry, F.: Grammaire comparée des langues classiques etc. Part 1. Phonétique (Paris *1868*).

Benfey, Th.: Geschichte der Sprachwissenschaft und orientalischen Philologie in Deutschland seit dem Anfange des 19. Jahrhunderts mit einem Rückblick auf die früheren Zeiten. Gesch. Wiss. Dtl., Neuere Zeit, vol. 8 (Munich *1869*).

Benn, G.: Goethe und die Naturwissenschaften, seine Kritik des Physikalismus. In G. Benn: Nach dem Nihilismus (Berlin *1932*).

Biermann, K.-R.: Aus der Entstehung der Fachsprache der Wahrscheinlichkeitsrechnung (with unpublished notes by C.F. Gauss). Forschungen und Forschritte *39:* 142–144 *(1965)*.

Bindseil, H.E.: Abhandlungen zur allgemeinen vergleichenden Sprachlehre. 1. Physiologie der Stimm- und Sprachlaute; 2. Über die verschiedenen Bezeichnungsweisen des Genus in den Sprachen (Hamburg *1838*).

Bischoff, E.: Über die phonetische Systematik des Sanskrit. Saunaka: Rig-Veda-Prātiśākhya. Vox *26:* 98–120 *(1916)*.

Bopp, F.: Vergleichende Grammatik des Sanskrit, Zend, Griechischen, Lateinischen, Litthauischen, Gothischen und Deutschen, 1st ed. (Berlin *1833–52*).

– Vergleichende Grammatik des Sanskrit, Send, Armenischen, Griechischen, Lateinischen, Altslavischen, Gothischen und Deutschen, 3rd ed. (Berlin *1868/1870/1871*).

Borinski, K.: Grundzüge des Systems der artikulierten Phonetik zur Revision der Principien der Sprachwissenschaft (Stuttgart *1891*).

v. Bortkiewicz, L.: Das Gesetz der kleinen Zahlen *(1898)*.

Bréal, M.: Essai de Sémantique – Science des significations. 1st ed. (Paris *1897*).

Breymann, H.: Die Phonetische Literatur von 1876–1895. Eine bibliographisch-kritische Übersicht (Leipzig *1897*).

de Brosches, Ch.: Traité de la formation méchanique des langues, et de principes physiques de l'étymologie. 2 vols. (Paris *1765*).

Brüche, E.: Drei Kapitel Galilei. 1. Gedanken um die Fallgesetze am Schiefen Turm zu Pisa; 2. Galileis Bild heute; 3. Galilei muss der "verdammten Lehre" abschwören. Physikal. Bl. *20: 65–80 (1964)*.

Brücke, E.: Phonetische Bemerkungen. Z. österr. Gymn. *8:* 749–768 *(1857)*.

– Grundzüge der Physiologie und Systematik der Sprachlaute für Linguisten und Taubstummenlehrer. 1st ed. (1856), 2nd ed. (Vienna *1876*).

Burckhardt, R.: Geschichte der Zoologie und ihrer wissenschaftlichen Probleme. 2nd ed., 2 vols. Samml. Göschen Nr. 357, 823 (Berlin/Leipzig *1921*).

Burdach, K.: Vom Mittelalter zur Reformation. Forschungen zur Geschichte der deutschen Bildung. 1st part (Halle *1893*).

– Deutsche Renaissance. Betrachtungen über unsere künftige Bildung (Berlin *1916*).

– Vorspiel. Gesammelte Schriften zur Geschichte des deutschen Geistes. Vol. 1, part 2: Reformation und Renaissance (Halle *1925*).

– Einleitung zum "Briefwechsel der Brüder Jacob und Wilhelm Grimm mit Karl Lachmann," vol. 1 (Jena *1927*).

– Die Wissenschaft von deutscher Sprache. Ihr Werden, ihr Weg, ihre Führer. 1st ed. (Berlin/Leipzig *1934*).

– Reformation, Renaissance, Humanismus. Zwei Abhandlungen über die Grundlage moderner Bildung und Sprachkunst. 1st ed. (1918), 3rd ed. (Wiss. Buchges., Darmstadt *1963*).

Cantor, G.: Grundlagen einer allgemeinen Mannigfaltigkeitslehre. Ein mathematisch-philosophischer Versuch in der Lehre des Unendlichen (Leipzig *1883*).

– Beiträge zur Begründung der transfiniten Mengenlehre. I and II. Math. Ann. *(1895/1897)*.

Cassirer, E.: Substanzbegriff und Funktionsbegriff, Untersuchungen über

die Grundfragen der Erkenntniskritik. 1st ed. (1910), 2d ed. (Berlin 1923).
– Philosophie der symbolischen Formen. 3 vols., part 1: Die Sprache. 1st ed. (1923), 2nd ed. (Wiss. Buchges., Darmstadt 1953).
– The Philosophy of Symbolic Forms: Vol. 1: Language. New Haven, Yale U.P., 1953.
(General) Catalogue of Printed Books. British Museum, Vol. 131 (1962).
Catalogue of Scientific Papers (1867–1902). Published by Royal Soc. of London. Vol. 3 (London (1869).
Champollion, J.F.: Lettre à M. Dacier . . . relative à l'alphabet des hiéroglyphes phonétiques employés par les égyptiens pour inscrire sur leurs monuments les titres, les noms et les surnoms des souverains grecs et romains (Paris 1822).
Charlier, C.V.L.: Vorlesungen über die Grundzüge der mathematischen Statistik. 2nd ed. (Lund 1920).
Court de Gébelin, A.: Monde primitif, analysé et comparé avec le monde moderne, consideré dans l'histoire naturelle de la parole; ou Origine du langage et de l'écriture (Paris 1775).
Crandall, I.B.: Theory of Vibrating Systems and Sound (New York 1926).
Curtius, E.R.: Europäische Literatur und lateinisches Mittelalter (Francke, Bern 1948).
Czuber, E.: Die statistischen Forschungsmethoden. 1st ed. (Vienna 1921).
Delbrück, B.: Einleitung in das Sprachstudium. Ein Beitrag zur Geschichte und Methodik der vergleichenden Sprachforschung. 3rd ed. Bibl. indog. Gramm., vol. 4 (Leipzig 1893).
Delitzsch, F.: Jesurun, sive Prolegomenon in Concordantias Veteris Testamenti. Vol. 3 (Grimma 1838).
Devrient, E.: Von der Entstehung des Stammbaums. Familiengeschichtl. Bl. 8: 2 (1910).
Dictionary of National Biography. Vol. 32 (London 1892).
Dictionnaire de l'Académie française (1878).
Diesterweg, F.A.W.: Beiträge zur Begründung der höheren Leselehre oder Anleitung zum lyrischen und euphonischen Lesen (Krefeld 1830).
Dieth, E.: Vademekum der Phonetik. 1st ed. (Francke, Bern 1950).
Dietrich A.J.: Kants Begriff des Ganzen in seiner Raum-Zeit-Lehre und das Verhältnis zu Leibniz. Abh. z. Philos. und ihrer Gesch., fasc. 50 (Halle 1916).
Diez, F.: Grammatik der romanischen Sprachen. Part 1 (Bonn 1836).
Dionysius Halicarnassensis: De compositione verborum. ed. G.H. Schäfer (London 1808).
Dodart, M.: Memoire sur les causes de la voix de l'homme et des ses différents tons. Hist. l'acad. roy. sci., année 1700 (1703).
Ehrentreich, A.: Zur Quantität der Tonvokale im Modern-Englischen. Auf Grund experimenteller Untersuchungen. Palaestra, Untersuchungen und Texte aus der deutschen und englischen Philologie, vol. 133 (Berlin 1920).
Engelhardt, V. and Gehrke, E.: Vokalstudien. Eine akustisch-psychologi-

sche Experimentaluntersuchung über Vokale, Worte und Sätze (Leipzig *1930*).

v. Essen, O.: Allgemeine und angewandte Phonetik. 1st ed. (1953), 3rd ed. (Akademie-Verlag, Berlin, *1962*).

Fechner, G.Th.: Elemente der Psychophysik. 2 vols., 1st ed. (Leipzig *1859/60*).

Ferrein, A.: De la formation de la voix de l'homme. Hist. l'acad. roy. sci., année 1741 *(1744)*.

Fischer, P.: Goethe-Wortschatz. Ein sprachgeschichtliches Wörterbuch zu Goethes sämtlichen Werken (Leipzig *1929*).

Fletcher, H.: Speech and Hearing (New York *1929*).

Föringer, O.: Lebensskizze Schmellers. Beilage zum 16. Jahresbericht des historischen Vereins Oberbayern (Munich *1855*).

Fónagy, I.: Die Metaphern in der Phonetik. Ein Beitrag zur Entwicklungsgeschichte des wissenschaftlichen Denkens. Janua Linguarum, Series Minor, No. 25 (Mouton, The Hague *1963*).

Fraenkel, A.: Einleitung in die Mengenlehre. 3rd ed. Grundlehr. math. Wiss., vol. 9 (Berlin *1928*).

Frank, E.: Mathematik und Musik und der griechische Geist. Logos *9: 222–259 (1920/21)*.

Friedenthal, R.: Goethe. Sein Leben und seine Zeit (Piper, München *1963*).

v. der Gabelentz, G.: Die Sprachwissenschaft, ihre Aufgaben, Methoden und bisherigen Ergebnisse. 1st ed. (1891), 2nd ed. (Leipzig *1901*).

Gamillscheg, E.: Etymologisches Wörterbuch der französischen Sprache (Heidelberg *1926–1929*).

Gemelli, A. und Pastori, G.: L'analisi elettroacustica del linguaggio. 2 vols. (Mailand *1934*).

Göppert, E.: Über die Herkunft des Wrisbergschen Knorpels. Ein Beitrag zur vergleichenden Anatomie des Säugethierkehlkopfs. Morph. Jb. *21: 68–149 (1894)*.

v. Goethe, J.W.: F.H. Jacobis auserlesener Briefwechsel, in zwey Bänden. Werke, vollständige Ausgabe letzter Hand, vol. 45; Nachgelassene Werke, vol. 5 (Stuttgart/Tübingen *1833*).

– Briefe an Friedrich August Wolf. ed. M. Bernays (Berlin *1868*).

– Wilhelm Meisters Wanderpahre. Werke (Sophie-Ausgabe), vol. 24 (Leipzig *1894*).

– Briefwechsel mit Wilhelm und Alexander von Humboldt. ed. L. Geiger (Berlin *1909*).

Gottheiner, V. und Zwirner, E.: Die Verwendung des Röntgentonfilms für die Sprachforschung. Fortschr. Geb. RöntgStrahl. *47: 455–462 (1933)*.

Grimm, J.: Deutsche Grammatik. Part 1, 3rd ed. (Göttingen *1840*).

Grimm, J. und W.: Briefwechsel der Brüder Jacob und Wilhelm Grimm mit Karl Lachmann. ed. A. Leitzmann. vol. 1 (Jena *1927*).

Gutzmann, H.: Röntgenfilmaufnahmen des Sprechens. Berichte I. Tagung Int. Ges. exp. Phon. (Bonn *1930*).

Hänle, C.H.: Praktische, zum Theil auf Musik gegründete Anleitung zur Declamation und zum mündlichen Vortrage (Frankfurt *1815*).

v. Haller, A.: Anfangsgründe der Physiologie des menschlichen Körpers (Elementa physiologiae corp. hum.). tr. J.S. Halle. vol. 3: Das Atemholen; Die Stimme (Berlin *1766*).

Hamburg, C.E.: Grundriß der körperlichen Beredtsamkeit (Bonn *1792*).

Hankamer, P.: Die Sprache, ihr Begriff und ihre Deutung im 16. und 17. Jahrhundert. Ein Beitrag zur Frage der literarhistorischen Gliederung des Zeitraums (Bonn *1927*).

Harnack, O.: Die Klassische Ästhetik der Deutschen. Würdigung der kunsttheoretischen Arbeiten Schillers, Goethes und ihrer Freunde (Leipzig *1892*).

Hatzfeld, A. und Darmesteter, A.: Dictionnaire général de la langue française du commencement du XVIIe siècle jusqu'à nos jours (Paris *1895*).

Hausdorff, F.: Mengenlehre. 2nd ed. Göschens Lehrbücherei, group 1, vol. 7 (Berlin *1927*).

Haym, R.: Wilhelm von Humboldt, Lebensbild und Charakteristik (Berlin *1856*).

van Helmont, F.M.: Kurtzer Entwurff des Eigentlichen Natur-Alphabets der Heiligen Sprache: Nach dessen Anleitung man auch Taubgebohrne verstehend und redend machen kann (Sulzbach 1667). Reprint by W. Viëtor) Vox *26*: Separatum=Phonet. Bibl. No. 1 *(1916)*.

Hellwag, Ch.F.: De formatione loquelae. thesis (Tübingen *1781*).

v. Helmholtz, H.: Über Goethes naturwissenschaftliche Arbeiten (1853). Vorträge und Reden, vol. 1, 4th ed. (Braunschweig *1896a*).

– Goethes Vorahnungen kommender naturwissenschaftlicher Ideen (1892). ibidem vol. 2 *(1896b)*.

– Über das Sehen des Menschen (1885). ibidem vol. 1 *(1896c)*.

Hertwig, R.: Lehrbuch der Zoologie. 1st ed. (1891), 11th ed. (Jena *1916*).

– A Manual of Zoology. tr. and ed. J. S. Kingsley (New York 1902).

Heuss, Th.: Justus von Liebig. Vom Genius der Forschung (Hamburg *1942*).

Hilbert, D.: Über den Zahlbegriff. Jahresbericht der deutschen Math.-Vereinig. *8:* 180–184 *(1900)*.

– Über die Grundlagen der Logik und der Arithmetik. proceedings III. Int. Math.-Kongr., Heidelberg 1904, pp. 174–185 *(1905)*.

– Grundlagen der Geometrie. 5. Aufl. Wiss. und Hypoth., vol. 7 (Leipzig/Berlin *1922*).

Hölder, H.: Geologie und Paläontologie. In Texten und ihrer Geschichte. Orbis Academicus, vol. II/11 (Alber, Freiburg/Munich *1960*).

Hönigswald, R.: Die Skepsis in Philosophie und Wissenschaft. Wege zur Philosophie, No. 7 (Göttingen *1914*).

– Zur Frage: Nichteuklidische Geometrien und Raumbestimmung durch Messung. Naturwissenschaften *3:* 307–311 *(1915)*.

– Die Philosophie von der Renaissance bis Kant. Geschichte der Philosophie, vol. 6 (Berlin/Leipzig *1923*).

– Immanuel Kant. Festrede (Breslau *1924*).

- Die Grundlagen der Denkpsychologie. Studien und Analysen. 1st ed. (1920), 2nd ed. (Leipzig *1925*).
- Grundfragen der Erkenntnistheorie. Kritisches und Systematisches. 1st ed. Beiträge zur Philosophie und ihrer Geschichte, vol. 1 (Tübingen *1931*).

Hofmann, K.: Andreas Schmeller. Eine Denkrede (Munich *1855*).

Hoppe, E.: Geschichte der Elektrizität (Leipzig *1884*).

- Mathematik und Astronomie im klassischen Altertum. Bibl. klass Altertumswiss., vol. 1 (Heidelberg *1911*).
- Das antike Weltbild. Arch. Gesch. Math. Naturw. Tech., vol. 5, fasc. 2 *(1913)*.
- Geschichte der Physik (Braunschweig *1926*).

v. Humboldt, W.: Ankündigung einer Schrift über die vaskische Sprache und Nation. Deutsches Museum *(1812)*.

- Die sprachphilosophischen Werke. Hrsg. und erklärt v. H. Steinthal (Berlin *(1883)*.
- Über die Verschiedenheit des menschlichen Sprachbaues und ihren Einfluß auf die geistige Entwicklung des Menschengeschlechts. 1st ed. (1836). Reprint (Berlin *1935*).
- Über das vergleichende Sprachstudium in Beziehung auf die verschiedenen Epochen der Sprachentwicklung (1820). Works, vol. 3 (Wiss. Buchges., Darmstadt *1963*).

Isserlin, M.: Die pathologische Physiologie der Sprache. Parts I and II. Ergebn. Physiol. *(1929/1931)*.

Jablonski, W.: Zum Einfluss der Goetheschen Farbenlehre auf die physiologische und psychologische Optik der Folgezeit. Arch. Gesch. Math. Naturw. Tech. *13: 75–81 (1930)*.

Jacobi, F.H.: Auserlesener Briefwechsel, vol. 2 (Leipzig *1827*).

Jacobi, J.G.: Sämmtliche Werke (4 vols.). vol. 1 (Zürich *1825*).

Jaensch, E.R.: Untersuchungen über Grundfragen der Akustik und Tonpsychologie (Leipzig *1929*).

Jellinek, A.: Die funktionelle Einordnung der Organismen in die Schallwelt. Arb. Wiener neurol. Inst. *34: 65–82 (1932)*.

Jespersen, O.: Fonetik, en systematisk fremstilling af lœren om sproglyd. 1st ed. (Copenhagen *1897–99*).

- Language, its nature, development, and origin. 1st ed. (London *1922*).
- Lehrbuch der Phonetik. tr. by H. Davidson. 1st ed. (1904), 4th ed. (Leipzig/Berlin *1926*).

Joachimi, M.: Die Weltanschauung der deutschen Romantik (Jena *1905*).

Johannsen, W.: Elemente der exakten Erblichkeitslehre. Mit Grundzügen der biologischen Variationsstatistik (tr.), 3rd ed. (Jena *1926*).

Junker, H.: Rede auf Wilhelm von Humboldt und die Sprachwissenschaft. Ber. Verh. sächs. Akad. Wiss. Leipzig. Phil.-Hist. Kl., vol. 87 (Leipzig *1936*).

Kalähne, A.: Grundzüge der mathematisch-physikalischen Akustik. 2 vols. (Leipzig/Berlin *1910/1913*).

Kant, I.: Kritik der reinen Vernunft (1781), 2nd ed. (Riga, *1787*).

- Kritik der Urteilskraft (1790). Werke (ed. E. Cassirer), vol. 5 (Berlin *1914*).
- Was heißt: sich im Denken orientieren? (1786). ibid. vol. 4 (Berlin *1922*).

Kayser, W.: Das sprachliche Kunstwerk. Eine Einführung in die Literaturwissenschaft (Francke, Bern *1948*).

v. Kempelen, W.: Mechanismus der menschlichen Sprache nebst der Beschreibung seiner sprechenden Maschine (Vienna *1791*).

Kennedy, A.G.: A Bibliography of Writings on the English Langauge etc. from the Beginning of Printing to the End of 1922 (Cambridge, Mass./ New Haven *1927*).

Kirby, W. und Spence, W.: An Introduction to Entomology: or, Elements of the Natural History of Insects. 4 vols., vol. 4 (London *1826*).

Klemm, O.: Wahrnehmungsanalyse. Hb. biol. Arbeitsmeth. Abt. 6: Meth. exp. Psychol. (Berlin/Vienna *1921*).

Klopstock, F.G.: Über die deutsche Rechtschreibung. Aus den "Fragmenten über Sprache und Dichtkunst" (Hamburg 1779). Sämmtliche sprachwissenschaftliche und ästhetische Schriften. ed. A.L. Back and A.R.C. Spindler. vol. 2 (Leipzig *1830a*).
- Über Etymologie und Aussprache. Aus den "Beyträgen zu der Hamburgischen Neuen Zeitung' (1781). ibidem (*1830b*).
- Grundsätze und Zweck unserer jetzigen Rechtschreibung. Aus dem "Musenalmanach von Voss und Gökingk für das Jahr 1782". ibidem (*1830c*).

Kneser, A.: Mathematik und Natur (Breslau *1918*).

Koch, F.: Herder und die Mystik. Bl. dt. Philosophie *1:* 1–29 *(1927/28)*.

Koch, M.: (Artikel über J.A. Schmeller). Der Wächter *6 (1923)*.

Köhler, L.: Die Melodie der Sprache in ihrer Anwendung besonders auf das Lied und die Oper. Mit Berührung verwandter Kunstfragen dargelegt (Leipzig *1853*).

Könnecke, G. Bilderatlas zur Geschichte der deutschen National-Literatur. Eine Ergänzung zu jeder deutschen Literaturgeschichte. 2nd ed. (Marburg *1895*).

Kohlbrugge, J.H.F.: Historisch-kritische Studien über Goethe als Naturforscher (Würzburg *1913*).
- Der Akademiestreit im Jahre 1831, der niemals enden wird. Biol. Zentralbl. *39:* 489–494 *(1919)*.

Kraft, V.: Erkenntnislehre (Springer, Vienna *1960*).

Krueger, F.: Beziehungen der experimentellen Phonetik zur Psychologie. Berichte II. Kongr. exp. Psychol., Würzburg 1906, pp. 58–122 *(1907)*.

Krumbacher, A.: Die Stimmbildung der Redner im Altertum bis auf die Zeit Quintilians. Rhet. Stud. Heft 10 (Paderborn *1921*).

Lammers, W.: Wilhelm von Humboldts Weg zur Sprachforschung 1785–1801. Neue dt. Forschung, Abtl. Sprachwissenschaft, vol. 1 (Berlin *1936*).

Latham, R.G.: Facts and Observations Relating to the Science of Phonetics.

Lond. Edinb. Dubl. Philos. Magaz. & Journ. of Sci. *18:* 124–130 *(1841a)*.
– The English language. 1st ed. (London *(1841b)*.
Latham, obituary for. The Athenaeum (London, 17 March *1888*).
Lauchert: M.J.W. Wocher. Allg. Dt. Biographie, vol. 43 (Leipzig *1898*).
v. Laue, M.: Erkenntnistheorie und Relativitätstheorie. Naturw. und Philos. Hrgs. v. G. Harig und J. Schleifstein (Akademie-Verlag, Berlin *1960*).
Lehnert, M.: Anfänge der wissenschaftlichen und praktischen Phonetik in England. Archiv. Stud. neuer. Spra. u. Lit. *174:28–35 (1938)*.
Leibniz, G.W.: (no title; concerning the Characteristica Universalis). Philosophische Schriften. ed. C.J. Gerhardt. vol. 7 (Berlin *1890*).
Lersch, L.: Die Sprachphilosophie der Alten. part 1: dargestellt an dem Streite über Analogie und Anomalie der Sprache. part 2: dargestellt an der historischen Entwicklung der Sprachkategorien. part 3: dargestellt an der Geschichte ihrer Etymologie (Bonn *(1838/1840/1841)*.
Lichtenberg, G.Ch.: Vermischte Schriften. ed. L.Ch. Lichtenberg and F. Kries. vol. 8 (Vienna *1844*).
Lischwitz, J.Ch.: Dissertatio de voce et loquela (Leipzig *1719*).
Löwi, M.: Schwellenuntersuchungen. Theorie und Experiment. Archiv. ges. Psychol. *48:* 1–73 *(1924)*.
Lorenz, O.: Lehrbuch der gesamten wissenschaftlichen Genealogie. Stammbaum und Ahnentafel in ihrer geschichtlichen, sociologischen und naturwissenschaftlichen Bedeutung (Berlin *1898*).
Lossen, J.: Anatomische Untersuchungen über die Cartilegines cuneiformes (Wrisbergsche Knorpel). thesis (Königsberg *1900*).
Luther, M.: Die gantze Heilige Schrift Deudsch. *(1533)*.
Menzerath, P. and de Lacerda, A.: Koartikulation, Steuerung und Lautabgrenzung. Eine experimentelle Untersuchung. Phonetische Studien (ed. P. Menzerath) Bd. 1 (Bonn/Berlin *1933*).
Merkel, C.L.: Rezension von E. Brücke: Grundzüge der Physiologie und Systematik der Sprachlaute (1856). Schmidts Jahrbuch ges. Med. *95:* 108–115 *(1857)*.
Merkel, C.L.: Über einige phonetische Streitpunkte. ibidem *100:* 86–101 *(1858)*.
– Die neueren Leistungen auf dem Gebiete der Laryngoskopie und Phonetik. Ebd. *108:* 81–103 *(1860)*.
– Anatomie und Physiologie des menschlichen Stimm- und Sprach-Organs (Anthropophonik). 1st ed. (1857), 2nd ed. (Leipzig *1863*).
– Physiologie der menschlichen Sprache (physiologische Laletik). (Leipzig *1866*).
Meyer, E.A.: Untersuchungen über Lautbildung. Festschrift Wilhelm Viëtor. Die Neueren Sprachen, supplementary volume, pp. 166–248 *(1910)*.
Meyer-Steineg, Th. and Sudhoff, K.: Geschichte der Medizin im Überblick mit Abbildungen. 1st ed. (1921), 2nd ed. (Jena *1922*).
Michaelis, Ch.F.: Die Kunst der rednerischen und theatralischen Declamation nach ältern und neuern Grundsätzen, über die Stimme, den Gesichtsausdruck aufgestellt usw. (Leipzig *1818*).

Miller, C.B.: Accent: Classes and Variations. Proc. II. Int. Congr. Phon. Sci. Lond. 1935 (London *1936*).

Müller, F.C.: Geschichte der organischen Naturwissenschaften im Neunzehnten Jahrhundert. Das Neunzehnte Jahrhundert in Deutschlands Entwicklung, ed. P. Schlenther, vol. 6 (Berlin *1902*).

Müller, J.: Untersuchungen über die Physiologie des Menschen *(1837a)*.

– Handbuch der Physiologie des Menschen für Vorlesungen. 2 vols, vol. 2, part 1 (Coblentz *1837b*).

– Über die Compensation der physischen Kräfte am menschlichen Stimmorgan. Mit Bemerkungen über die Stimme der Säugetiere, Vögel und Amphibien. Fortsetzung und Supplement über die Physiologie der Stimme (Berlin *1839*).

Müller, M.: Rig-Veda-Prātiśākhya, das älteste Lehrbuch der vedischen Phonetik. Sanskrittext mit Übersetzung und Anmerkungen (Leipzig *1869*).

Murray, J. and others: New English Dictionary, on Historical Principles (Oxford *1884–1933*).

Nadler, J.: Literaturgeschichte der deutschen Stämme und Landschaften. 1st ed. (1912), 2nd ed., vol. Bd. 3 (Regensburg *1923*).

Nagel, W.: Die Wirkungen des Lichtes auf die Netzhaut. Handbuch der Physiologie des Menschen, vol. 3, 2nd part (Braunschweig *1905*).

Narcissus, Lord Bishop of Ferns and Leighlin: An Introductory Essay of the Doctrine of Sounds, Containing some Proposals for the Improvement of Acoustics. Phil. Trans., vol. 14 (Oxford *1684*).

Neumann, F.: Die Taubstummen-Anstalt zu Paris im Jahre 1822. Eine historisch-pädagogische Skizze (Königsberg *1827*).

Newton, I.: Philosophiae Naturalis Principia Mathematica. 1. Aufl. (London *1687*).

Nicklas, J.: Johann Andreas Schmellers Leben und Wirken: Eine Festgabe zum 100 jährigen Geburtstage des großen Sprachforschers (Munich *1885*).

Oksaar, E.: Semantische Studien im Sinnbereich der Schnelligkeit. "Plötzlich", "schnell" und ihre Synonymik im Deutsch der Gegenwart und des Früh-, Hoch- und Spätmittelalters. Acta Universitatis Stockholmiensis, Stockholmer Germanistische Forschungen, 2 (Almquist & Wiksell, Stockholm *1958*).

Ostwald, W.: Einführung in die Farbenlehre. Reclams Univ. Bibl. 6041–6044 (Leipzig *1919*).

– Die Harmonie der Formen (Leipzig *1922*).

– Farbkunde. Ein Hilfsbuch für Chemiker, Physiker, Naturforscher usw. Chemie und Technik der Gegenwart, vol. 1 (Leipzig *1923*).

Otto, E.: Grundfragen der Linguistik. Indog. Forschung. *52:* 177–195 *(1934)*.

Panconcelli-Calzia, G.: Zur Geschichte der phonoskopischen Vorrichtungen. Ann. Physik, 5. series *10:* 673–680 *(1931)*.

Panconcelli-Calzia, G.: Das Hören durch die Zähne. Eine phonetisch-geschichtliche Vorstudie. Med. Welt. *23:* 1–12 *(1934)*.

184   BIBLIOGRAPHY

- Zur Geschichte des Kymographions. Z. Laryng. Rhinol. Otol. u. ihre Grenzgeb. *3:* 196–207 *(1935a)*.
- Der erste Kehlkopfspiegel: Babingtons "Glottiskop" (1829–1835). Med. Welt. *48:* 1–19 *(1935b)*.

Paneth, F.: Über die erkenntnistheoretische Stellung des chemischen Elementbegriffs. Lecture (Halle *1931*).

Paul, H.: Geschichte der germanischen Philologie. Grundriss der germanischen Philologie, ed. H. Paul, vol. 1. 1st ed. (Strassburg *1891*).
- Prinzipien der Sprachgeschichte. 1st ed. (1880), 5th ed. (Halle *1920*).
- Principles of the History of Language. Sonnenschein (London *1881*).

Pilch, H.: Phonemtheorie, part 1. Bibl. Phonetica, Fasc. 1 (Karger, Basel/ New York *1964*).

Pitman, I.: A history of phonography — How it came about (lecture). Report in Phonetic Journal, 1 August *(1868)*.

Porzig, W.: Das Wunder der Sprache. 2nd ed. (Francke, Bern *1957*).

Pott, A.F.: Etymologische Späne. Z. vergl. Sprachforsch. *5:* 241–300 *(1856)*.

Quetelet, L.A.: Anthropométrie, ou mesure des différentes facultés de l'homme. (Brussels *1870*).

Rádl, E.: Geschichte der biologischen Theorien seit dem Ende des 17. Jahrhunderts. part 1 (Leipzig *1905*).

Rapp, K.M.: Versuch einer Physiologie der Sprache, nebst historischer Entwicklung der abendländischen Idiome nach physiologischen Grundsätzen. vol. 1: Die vergleichende Grammatik als Naturlehre dargestellt. Erster oder physiologischer Theil (Stuttgart/Tübingen *1836*).

v. Raumer, R.: Die Aspiration und die Lautverschiebung. Eine sprachgeschichtliche Untersuchung. 1st ed. (1837). Gesammelte sprachwiss. Schriften. ed. Heyder and Zimmer (Frankfurt/Erlangen *1863a*).
- Offener Brief an den Herausgeber der Zeitschrift für die deutschen Mundarten, 1857. ibid. *(1863b)*.
- Geschichte de germanischen Philologie, vorzugsweise in Deutschland. Geschichte der Wissenschaften in Deutschland, Neuere Zeit, vol. 9 Munich *1870*).

Renouf, P. le Page: Lectures on the Origin and Growth of Religion as Illustrated by the Religion of Ancient Egypt (delivered 1879). (London *1880*).

Richert, G.: Die Anfänge der romanischen Philologie und die deutsche Romantik. Beitr. Gesch. roman. Sprachen u. Lit., ed. F. Mann, vol. 10 (Halle *1914*).

Rickert, H.: Das Eine, die Einheit und die Eins. Bemerkungen zur Logik des Zahlenbegriffs (1911). 2nd ed. (Tübingen *1924*).

Riehl, A.: Der philosophische Kritizismus. Geschichte und System. Bd. 1: Geschichte des philosophischen Kritizismus. 1st ed. (Leipzig *1876*).
- Galileo Galilei. In A. Riehl: Führende Denker und Forscher (Leipzig *1922*).

Riemann, B.: Über die Hypothesen, welche der Geometrie zu Grunde liegen (inaugural lecture, Göttingen 1854). (Berlin *1919*).

Ritter, G.: Studien zur Spätscholastik. *I:* Marsilius von Inghen und die okkamistische Schule in Deutschland. *II:* Via antiqua und via moderna auf den deutschen Universitäten des XV. Jahrhunderts. Sitzungsbericht Heidelberger Akad. d. Wiss. Heidelberger universitätsgesch. Forschung. *(1921/22).*

Rockinger, L.: An der Wiege der baierischen Mundart-Grammatik und des baierischen Wörterbuches. Oberbayer. Arch. vaterländ. Gesch. *43 (1886).*

v. Rohr, M.: Das photographische Objektiv (Die Perspektive der Aufnahme; die Abbildungstiefe photographischer Aufnahmen). Hb. Physik. *18:* 419–453 (Berlin *1927*).

v. Rohr, M.: Die Strahlenbegrenzung in ihrer Bedeutung für die Lichtbilder. Hb. wiss. angew. Photographie, vol. 1 (Vienna *1932*).

Rostovtzeff, M.: Geschichte der Alten Welt. tr. by H.H. Schaeder. vol. 1: Der Orient und Griechenland. Sammlung Dieterich, vol. 72 (Leipzig *1941*).

Rousselot, P.-J.: Principes de phonétique expérimentale. 1st ed. (1897/ 1909). 2 vols. (Paris/Leipzig *1924/1925*).

Runge, J.G.: Dissertatio inaug. de voce eiusque organis (Lyon *1753*).

Russell, M.: View of Ancient and Modern Egypt with an Outline of its Natural History (Edinburgh *1831*).

– Egypt XI. *(1853).*

Salt, H.: Essay on Dr. Young's and Champollion's phonetic system of hieroglyphics (London *1825*).

de Saussure, F.: Grundfragen der allgemeinen Sprachwissenschaft (Cours de linguistique générale. ed. Ch. Bally und A. Sechehaye. 1st ed. Lausanne/Paris 1916). tr. by H. Lommel (Berlin/Leipzig *1931*).

Schaeder, J.A.: Zur Schallstärke- und Lautstärke-Registrierung für phonometrische Zwecke. Arch. vergl. Phonetik *2:* 235–245 *(1939).*

Schaefer, K.: Psychologische Akustik. Hb. biol. Arbeitsmeth., Abt. 6: Meth. exp. Psychol. (Berlin/Vienna *1922*).

Schelhamer, G.Ch.: De voce eiusque effectibus. thesis (Jena *1677*).

Scherer, W.: Zur Geschichte der deutschen Sprache (1868), 2nd ed. (Berlin *1878*).

– Zur Regelung der deutschen Rechtschreibung (1869). Kleine Schriften, ed. K. Burdach and E. Schmitt. vol. 1, pp. 418–426 (Berlin *1893a*).

– Rezension von "Grundzüge der Physiologie und Systematik der Sprachlaute" von Ernst Brücke (2nd ed. 1876). ibid. pp. 268–275 (Berlin *1893b*).

– Jacob Grimm. (Reprint of the 2nd ed. by S. v. d. Schulenburg). Der Domschatz, vol. 9 (Berlin *1921*).

Schiller, F.: Briefe. ed. F. Jonas. Kritische Gesamtausgabe. vol. 3 (Stuttgart/ Leipzig/Berlin/Vienna *1893*).

Schlegel, F.: Über die Sprache und die Weisheit der Indier. Ein Beitrag zur Begündung der Altertumskunde (Heidelberg *1808*).

Schocher, Ch.G.: Soll die Rede auf immer ein dunkler Gesang bleiben, und können ihre Arten, Gänge und Beugungen nicht anschaulich gemacht und nach der Art der Tonkunst gezeichnet werden? thesis (Leipzig *1792*).

Schole, H.: Experimentelle Untersuchungen an höchsten und an kürzesten Tönen. (Zugleich ein Beitrag zur Theorie der akustischen Elementarphänomene). Z. Psychol. *131 (1934)*.

Scholtz, M.: Lehrbuch der pharmazeutischen Chemie. vol. 1: Anorganischer Teil (Heildelberg *1910*).

Schröder, E.: J.A. Schmeller. Allg. Deutsche Biographie, vol. 31 (Leipzig *1890*).

H. Schuchardt-Brevier. Ein Vademecum der allgemeinen Sprachwissenschaft. ed. L. Spitzer. 1st ed. (1922), 2nd ed. (Halle *1928*).

Schuster, J.: Oken, Welt und Wesen, Werk und Wirkung. Arch. Gesch. Math. Naturw. Tech. *12*:54ff. *(1929)*.

– Die Anfänge der wissenschaftlichen Enforschung der Geschichte des Lebens durch Cuvier und Geoffroy Saint-Hilaire. Eine historisch-kritische Untersuchung. ibidem *12*: 269–336, 341–386; *13*: 1–64 *(1930/ 31)*.

Scripture, E.W.: Review by P.W. Bridgman, The logic of physics (1928). Z. Exp.-Phonetik *1*: 171–172 *(1930/31a)*.

– Review by H. Gutzmann, Röntgenfilmaufnahmen des Sprechens (1930). ibidem pp. 172–173 *(1930/31b)*.

Sheridan, Th.: A course of Lectures on Elocution: together with Two Dissertations on Language and Some Other Tracts Relative to these Subjects (London *1762*).

Sheridan, Th.: Lectures on the Art of Reading. 2 vols. (London *1775*).

– Über die Declamation oder den mündlichen Vortrag in der Prose und Poesie. tr. by R.C. Lobel (Leipzig *1792*).

Sievers, E.: Grundzüge der Lautphysiologie, zur Einführung in das Studium der Lautlehre der indogermanischen Sprachen. 1st ed. (Leipzig *1876*), 2nd ed.: Grundzüge der Phonetik zur Einführung usw. Bibl. indogerm. Grammatiken, vol. 1 (Leipzig *1881*).

Standelmann, F.: Erziehung und Unterricht bei den Griechen und Römern (Trieste *1891*).

Staiger, E.: Die Zeit als Einbildungskraft des Dichters. Untersuchungen zu Gedichten von Brentano, Goethe und Keller (Zürich *1939*).

– Grundbegriffe der Poetik (Atlantis, Zürich *1946*).

– Die Kunst der Interpretation. Studien zur deutschen Literaturgeschichte (Atlantis, Zürich *1955*).

Steinthal, H.: Geschichte der Sprachwissenschaft bei den Griechen und Römern mit besonderer Rücksicht auf die Logik. 2 parts. 2nd ed. (Berlin *1890/91*).

Stenzel, J.: Über den Einfluß der griechischen Sprache auf die philosophische Begriffsbildung. Neue Jahrbücher klass. Altertum u. Pädag. *(1921)*.

– Sinn, Bedeutung, Begriff, Definition. Ein Beitrag zur Frage der Sprachmelodie. Jahrbuch Philol. *1*: 160–201 *(1925)*.

– Philosophie der Sprache (Munich/Berlin *1934*).

Storr, G.C.Ch.: De formatione loquelae (Tübingen *1781*).

Stumpf, K.: Die Sprachlaute. Experimentell-phonetische Untersuchungen nebst einem Anhang über Instrumenterklärungen (Berlin *1926*).

Stutterheim, C.F.P.: Wijsbegeerte en Taalwetenschap. Problemen der Fonologie. Alg. Nederl. T. wijsbeg. en Psychol. *1:* 218ff. *(1959)*.

Süssmilch, J.P.: Versuch eines Beweises, daß die erste Sprache ihren Ursprung nicht vom Menschen, sondern von Gott erhalten habe (Berlin *1766*).

Sulze, W.: Die physikalische Analyse der Stimm- und Sprachlaute. Hb. norm. u. path. Physiol., 2nd ed., vol. 15 (Berlin *1931*).

Swoboda, H.: Der Schachroboter. Herr von Kempelen und sein Automat. Frankfurter Allg. Zeitung (6 March *1965*).

v. Thiersch, F.: Gedächtnisrede auf Schmeller. Sitzungsbericht Bayr. Akad. der Wiss. *(1852)*.

Thomsen, V.: Geschichte der Sprachwissenschaft bis zum Ausgang des 19. Jahrhunderts. Kurzgefaßte Darstellung der Hauptpunkte. tr. by H. Pollack (Halle *1927*).

Thürnagel, E.: Systematische Anleitung zur Declamation, für Jeden, dessen Beruf ein gründliches Studium derselben erfordert (Heidelberg *1825*).

– Theorie der Schauspielkunst (Heidelberg *1836*).

Tönnies, J.F.: Der Neurograph, ein Apparat zur Aufzeichnung bioelektrischer Vorgänge unter Ausschaltung der photographischen Kurvendarstellung. Naturw. *20:* 381–387 *(1932)*.

Trendelenburg, F. (ed.): Akustik. Hb. der Physik. ed. H. Geiger and K. Scheel. vol. 8 (Berlin *1929*).

– Ohms akustisches Gesetz. Akust. Z. *4:* 88–89 *(1939)*.

Trubetzkoy, N.S.: Grundzüge der Phonologie. 1st ed. (1939), 2nd ed. (Vandenhoeck & Ruprecht, Göttingen *1958a*).

– Anleitung zu phonologischen Beschreibungen. 1st ed. (1935), 2nd ed. Lautbibliothek der deutschen Mundarten, fasc. 2 (Vandenhoeck & Ruprecht *1958b*).

– Introduction to the Principles of Phonological Descriptions. tr. by L.A. Murray, ed. H. Bluhme. Nijhoff, The Hague, 1968.

Viëtor, W.: Elemente der Phonetik des Deutschen, Englischen und Französischen. 1st ed. (1884), 7th ed. (Leipzig *1923*).

Volkmann, R.: Die Rhetorik der Griechen und Römer, in systematischer Übersicht dargestellt. 2nd ed. (Leipzig *1885*).

Vorländer, K.: Kant, Schiller, Goethe. Gesammelte Aufsätze (Leipzig *1907*).

Wagner, R.: Samuel Thomas Sömmerings Leben und Verkehr mit seinen Zeitgenossen. In: Vom Bau des menschlichen Körpers (v. Sömmering), vol. 1 (Leipzig *1844*).

Wallis, J.: De loquela, sive sonorum formatione, tractus grammaticophysicus *(1653)*.

v. Wartburg, W.: Einführung in Problematik und Methodik der Sprachwissenschaft. 1st ed. (1943), 2nd ed. (Niemeyer, Tübingen *1962*).

Watts, Th.: Dr. R. G. Latham. The Athenaeum, No. 3151: 340–341 (17 March *1888*).

Weber, H.H.: Zur Frage der natürlichen Anordnung der Farben im Farbenkreis. Forschungen und Fortschritte *14:* 186–188 *(1938)*.

Weiss, O.: Stimmapparat des Menschen. Handbuch norm. u. path. Physiol., vol. 15, 2nd part (Berlin *1931*).

Welcker, F.G.: Zoegas Leben. Sammlung seiner Briefe und Beurtheilung seiner Werke. 1st ed. (1819). Reprint: Klassiker der Archäologie, vols. 2 and 4 (Halle *1912/13*).

Westphal, R.: Aristoxenos von Tarent, Melik und Rhythmik des classischen Hellenenthums. 2 parts (Leipzig *1883/1893*).

– Griechische Metrik (Leipzig *1887*).

Weyl, R.: Die geologischen Studien Leonardo da Vincis. Philosophia naturalis *1 (1950)*.

Winteler, J.: Die Kerenzer Mundart des Kantons Glarus in ihren Grundzügen dargestellt (Leipzig/Heidelberg *1876*).

Wocher, M.: Allgemeine Phonologie, oder natürliche Grammatik der menschlichen Sprache. Mit specieller Anwendung auf das Hebräische, Griechische, Lateinische, Italienische, Französische, Englische, Deutsche und die resp. alten und neuen Mundarten (Stuttgart/Tübingen *1841*).

Wolandt, G.: Die moderne Grundlegungstheorie und der Galileische Naturbegriff. Philosophia naturalis *8:* 191–197 *(1964)*.

Wolff, W.: Selbstbeurteilung und Fremdbeurteilung im wissentlichen und unwissentlichen Versuch. Physiognomische Untersuchungen an der Stimme, dem Profil, den Händen und einer freien Nacherzählung. Psychol Forsch. *(1932)*.

Zipf, G.K.: Phonometry, Phonology, and Dynamic Philology: an Attempted Synthesis. American Speech *13:* 275–285 *(1938)*.

Zoega, G.: De origine et usu Obeliscorum (Rom *1797*).

Zwirner, E.: Zum Begriff der Geschichte. Eine Untersuchung über die Beziehungen der theoretischen zur praktischen Philosophie (Leipzig *1926*).

– Quantität, Lautdauerschätzung und Lautkurvenmessung (Theorie und Material). Arch. Néerl. de Phon. Exp. *8/9:* 236–246 *(1933a)*.

– Gestikulationskurven. ibid. 278–279 *(1933b)*.

– Phonetische Tonhöhenbezifferung. ibid. 284–289 *(1933c)*.

– Die Grundlagen der vergleichenden Sprachphysiologie und Sprachphysik. part 1: Passow-Schaefer Beitr. vol. *31:*148–180 *(1934)*.

– Textliste märkischer Mundart. Phonometrische Forschungen, series B, vol. 2 (Berlin *(1936a)*.

– Textliste schlesischer Mundart. Phonometrische Forschungen, series B, vol. 3 (Berlin *(1936b)*.

– Textliste neuhochdeutscher Vorlesesprache bayrischer Färbung. Phonometrische Forschungen, series B, vol. 5 (Berlin *1937*).

– Langue et parole en phonométrie. Annu. Inst. Philol. et Hist. Orient. et Slav. *6:* 391–394 *(1938)*.

– Phonologie und Phonetik. Acta Ling. Rev. Intern. *1:* 29–47 *(1939)*.

– Das Gespräch. Beitrag zur Theorie der Sprache und der universitas litterarum. Studium Generale *4:* 213–227 *(1951)*.

– Die Konsultation. Zur Theorie der psychophysischen Korrelationen und der "psychosomatischen Ganzheit". Schweiz. med. Wschr. *83:* 1512–1529 *(1953)*.

Zwirner, E.: Anleitung zu sprachwissenschaftlichen Tonbandaufnahmen. Lautbibliothek deutscher Mundarten, fasc. 31 (Vandenhoeck und Ruprecht, Göttingen *1964a*).

– Die Beziehungen der Phonemtheorie Trubetzkoys zur Phonetik. Wiener physik. part 1: Passow-Schaefer Beitr. vol. *31:* 148–180 *(1934)*.

– Die Bedeutung der Sprachstruktur für die Analyse des Sprechens. Problemgeschichtliche Erörterung. Proc. V. Intern. Congr. Phonet. Sci. Münster 1964 (Karger, Basel/New York *1965*).

Zwirner, E., Maack, A. and Bethge, W.: Vergleichende Untersuchungen über konstitutive Faktoren deutscher Mundarten. Z. Phonetik *9:* 14–30 *(1956)*.

Zwirner, E. and Richter, H. (ed.): Gesprochene Sprache – Probleme ihrer strukturalistischen Untersuchung. Forschungsberichte der Deutschen Forschungsgemeinschaft No. 7. (Steiner, Wiesbaden *1966*).

Zwirner, E. and Zwirner, K.: Lauthäufigkeit und Zufallsgesetz. Forschungen und Fortschritte *11:* 43–44 *(1935a)*.

– Phonometrischer Beitrag zur Frage der neuhochdeutschen Lautmelodie. Vox *21:* 45–70 *(1935b)*.

– Textliste neuhochdeutschen Vorlesesprache schlesischer Färbung. Phonometrische Forschungen, series B, vol. 1 (Berlin *1936a*).

– Phonometrischer Beitrag zur Frage des neuhochdeutschen Akzents. Indogermanische Forschungen *54:*1–32 *(1936b)*.

– Streuung sprachlicher Merkmale. Forschungen und Fortschritte *12:* 191–192 *(1936c)*.

– Lesebuch neuhochdeutscher Texte. Phonometr. Forschungen, series B, vol. 4 (Berlin *1937a*).

– Phonometrischer Beitrag zur Frage der neuhochdeutschen Quantität. Archiv vergl. Phon. *1:* 96–113 *(1937b)*.

Zwirner, K.: Das Eindringen statistischer Forschungsmethoden in die Sprachvergleichung. Archiv vergl. Phon. *1:* 116–120 *(1937)*.

Index of Persons

Amman, J. C., 30, 32, 156
Anaxagoras, 17
Arens, H., 25
Aristotle, 15–18, 35, 36
Aristoxenus of Tarentum, 17
Attwood, G., 24
Azoulay, 74

Bacon of Verulam, Francis, 30
Baer, Karl Ernst von, 46
Baier, J. J., 54
Baker, C., 31
Barlow, 14
Baudouin de Courtenay, J., 73, 77–79, 81
Baudry, F., 8
Behaghel, O., 74
Bell, A. G., 14
Benfey, Thomas, 15, 18
Berliner, E., 74
Bernoulli, J., 30
Bindseil, H. E., 11
Blainville, H. M., 41
Blumenbach, J. F., 46
Bodmer, J. J., 38
Böhme, Jacob, 26
Böhtlingk, Otto von, 15
Boerhaave, H., 34
Bonet, J. P., 27
Bopp, F., 9, 11, 15, 19, 20, 55, 56, 58–60, 62, 75
Bortkiewicz, L. von, 154
Bréal, M., 8f
Brenner, J., 33
Broca, P., 55
Brosches, Charles de, 32
Brücke, E. W., 13–15, 18, 60, 64, 65, 81

Brugmann, F. K., 64
Bruno, G., 48
Bühler, K., 78, 104
Buffon, G. L. Leclerc, Comte de, 41
Burdach, K., 9, 18, 38, 39

Carnap, R., 104
Cassirer, E., 15, 19, 24f, 26, 39, 57, 58
Catherine II, 52
Champollion, J. F., 8, 9, 10
Charlier, C. V. L., 149
Chladni, E. F. F., 5
Coerdoux, 51, 52
Copernicus, 48
Court de Gébelin, A., 37
Couturat, L., 27f
Crandall, I. B., 71
Curtius, D. R., 19, 56
Cusanus, N., 20
Cuvier, G. D., 40–42, 46, 55, 74
Czuber, E., 53

Delgarno, G., 27
Dante Alighieri, 43, 126
Delbrück, B., 18, 56, 75
Delitzsch, F., 10
Democritus, 17
Descartes, René, 26, 28–29
Devrient, E., 38
Diesterweg, F. A. W., 30
Diez, F., 9, 10, 55
Dionysius of Halicarnassus, 17
Dodart, M., 32
Dürr, K., 104
Dulong, P. L., 60

Edison, Thomas A., 73, 74, 77
Eichhorn, J. A. F., 36
Ellis, A. J., 13, 14

Epicurus, 17
Euler, L., 5
Ewing, J. A., 74
Exner, F., 74
Exner, S., 74

Fabricius ab Aquapendente, H., 36
Ferrein, A., 32
Fletcher, H., 71
Fourier, J. B. J., 5, 70
Frank, E., 17
Frommann, G. K., 65
Fulda, F. K., 27

Gabelentz, H. G. Conon von der, 58,
    75, 76, 81, 123, 128
Galen, 32, 34
Galileo Galilei, 20–25, 28, 43, 48, 50
Galton, Sir F., 47
Gauss, C. F., 65, 154
Gay-Lussac, J. L., 55
Geiger, H., 71
Geoffroy Saint-Hilaire, 42, 55
Gervinus, G. G., 50
Goethe, Johann Wolfgang von, 33,
    39, 41–47, 51, 55, 57
Goeze, J. M., 36
Gottheiner, V., 107, 110, 111
Gravesande, G.-J., 30
Grimm, J., 11, 20, 50, 56, 58, 61, 75, 81
Grimm, W., 50, 55, 62, 63
Gutzmann, H., 107, 110, 111
Gyármath, S., 52

Hänle, C. H., 30
Haller, Albrecht von, 34, 35, 37, 61
Hamann, J. G., 39
Hamilton, Sir William, 53
Hartel, Wilhelm von, 74
Harvey, W., 34
Havet, L., 14
Hegel, G. W. F., 19
Heinzel, R., 74
Hellwag, C. F., 32
Helmholtz, H. von, 5, 49, 68, 71
Helmont, F. M. van, 28, 37
Helmont, J. B. van, 28
Heraclitus, 15
Herder, J. G., 39, 44, 48, 50, 51, 57, 61
Hermann, F. B. W., von, 74
Hermann, L., 5, 65
Hervás y Panduro, L., 53

Hildebrand, R., 38
Hönigswald, R., 21
Hooke, R., 41
Hoppe, E., 30
Humboldt, Wilhelm von, 19, 47, 48,
    50, 51, 55–59, 63

Isidore of Seville, 38

Jacobi, F. H., 43
Jagić, Vatroslav, 74
Jenkin, Fleeming, 74
Jespersen, Otto, 33, 86, 126
Johannsen, W., 47
Jones, W., 52, 53

Kalähne, A., 71
Kant, Immanuel, 20, 43–51, 57, 63
Kekule von Stradonitz, 74
Kempelen, Wolfgang von, 11, 37, 39,
    60–62
Kepler, J., 48
Kirby, W., 9
Klopstock, F. G., 9
Kluge, F., 74
Knebel, K. L. von, 44
Köppen, 43
Körner, Christian G., 44, 47, 48, 57
Kräuter, J. F., 81
Kraft, V., 104
Kraus, C. J., 52
Krumbacher, A., 15

Lachmann, K., 59, 63
Lahr, J., 74
Lamarck, J. B. de Monet de, 41, 46
Langle, 53
Latham, R. G., 11, 13, 17
Laue, Max von, 80
Leibniz, G. W., 26, 28–29, 32
Leonardo da Vinci, 41, 54
Lersch, L., 15, 17
Leskien, A., 64
Lichtenberg, G. C., 33
Liebeg, 55
Linnaeus, 35, 36, 39
Ludwig, C. F. W., 14
Luther, Martin, 38

Malpighi, M., 36
Massmann, H. F., 64
Mayer, T., 33

Meckel, J. F., 37
Meinhof, C., 137
Mendel, G., 47
Menzerath, P., 84
Merkel, C. L., 13, 14, 30, 61, 63
Mersenne, M., 26, 30
Mertrud, 55
Meyer, E. A., 106–9
Meyer, H. von, 55
Michaelis, C. F., 30
Morgagni, G. B., 35
Müller, J., 11
Müller, K., 74
Müller, M., 15

Narcissus, 29
Natorp, P., 22f
Newton, Isaac, 30, 43, 61
Niebuhr, B. G., 50

Ohm, G. S., 5, 68–71
Oken, L., 37, 43
Orbigny, A. Dessalines d', 42

Pallas, P. S., 52
Pap, A., 104
Paracelsus (T. B. von Hohenheim), 34
Pascal, B., 153
Paul, H., 18, 73, 75, 80, 85
Pearson, K., 47
Peirce, C., 104
Penttilä, A., 104
Petit, A. T., 60
Plato, 15–17, 20, 23, 48, 50
Ponce, P., 27
Pott, A. F., 9, 19, 56
Pythagoras, 17

Quetelet, L. A., 47, 80, 155, 156

Rameau, J.-P., 68
Ramus, P., 19
Rapp, K. M., 10, 11, 62, 63
Rask, R., 20, 55, 58, 61
Raumer, K. von, 64
Raumer, R. von, 4, 10, 13, 18, 64, 65,
    67, 73, 77, 80
Ray, 35
Reichenbach, H., 104
Riehl, A., 22f, 24f
Ritter, G., 20
Robinson, 26

Rosapelly, C. L., 14
Rousselot, P.-J., 7, 8, 18, 85
Runge, J. G., 32
Russell, M., 9f

Saarnio, 104
Salt, H., 8
Sassetti, F., 25, 26, 51, 52
Saussure, F. de, 58, 75, 76, 78, 79,
    81, 128
Savigny, F. K. von, 50
Schäffer, G. A., 36
Scheel, K., 71
Schelhamer, G. C., 29
Schelling, F. W. J. von, 39, 64
Scherer, W., 13, 18, 47, 60, 64, 65
Schiller, F. von, 44, 48, 50, 51
Schlegel, F. von, 53, 54, 55, 56, 59
Schleicher, A., 19
Schmeller, A., 55, 63, 64
Schocher, C. G., 30
Schröder, E., 64
Scripture, E. W., 80, 82, 83, 107–10,
    132, 133
Seebeck, A., 5, 67, 69–71
Sheridan, Thomas, 30
Sievers, E., 14, 18, 32, 97
Sommer, R., 74, 77
Spence, W., 9
Stadelmann, F., 15
Stein, Baroness Charlotte von, 44
Steinthal, H., 15
Steno, N., 54
Stenzel, J., 15, 16
Storr, G. C. C., 32
Stumpf, C., 5, 74
Stutterheim, C. F. P., 80
Süssmilch, J. P., 65
Sweet, H., 14

Techmer, F., 74
Thürnagel, E., 30
Tönnies, J. F., 148
Trendelenburg, F., 5, 71
Trubetzkoy, N. S., 78, 126, 133

Usener, H., 5

Valla, L., 19
Venzky, G., 32
Vesalius, Andreas, 34
Vico, G. B., 39

Viëtor, W., 32, 81, 106
Vives, L., 19

Wagner, K. W., 5
Waller, R., 33
Wallis, J., 26, 28
Wenker, G., 64, 67, 73, 74
Werkmeister, 30
Wernicke, C., 55
Wilkins, J., 27
Willis, R., 5
Winckelmann, J. J., 50
Windelband, W., 23f

Winteler, J., 73, 77
Wocher, M., 11
Wolf, F. A., 48, 50, 51, 57
Wolff, C. F., 32, 37, 46
Wotton, E., 35
Wrede, F., 64, 74
Wrisberg, H. A., 35

Young, Thomas, 8

Zarlino, G., 30
Zoega, G., 8f, 10